Contemporary Issues in Technology Education

Series Editors
P John Williams
Curtin University, Perth, Australia

Alister Jones
University of Waikato, Hamilton, New Zealand

Cathy Buntting
University of Waikato, Hamilton, New Zealand

Marc de Vries
Technische Universiteit Delft, Delft, The Netherlands

Contemporary Issues in Technology Education - About this series

Technology education is a developing field, new issues keep arising and timely, relevant research is continually being conducted. The aim of this series is to draw on the latest research to focus on contemporary issues, create debate and push the boundaries in order to expand the field of technology education and explore new paradigms. Maybe more than any other subject, technology education has strong links with other learning areas, including the humanities and the sciences, and exploring these boundaries and the gaps between them will be a focus of this series. Much of the literature from other disciplines has applicability to technology education, and harnessing this diversity of research and ideas with a focus on technology will strengthen the field.

More information about this series at http://www.springer.com/series/13336

P John Williams • David Barlex
Editors

Explorations in Technology Education Research

Helping Teachers Develop Research Informed Practice

Springer

Editors
P John Williams
Curtin University
Perth, Australia

David Barlex
University of Exeter Associate
Brighton, UK

ISSN 2510-0327 ISSN 2510-0335 (electronic)
Contemporary Issues in Technology Education
ISBN 978-981-13-3009-4 ISBN 978-981-13-3010-0 (eBook)
https://doi.org/10.1007/978-981-13-3010-0

Library of Congress Control Number: 2018967693

This Springer imprint is published by the registered company Springer Nature Singapore Pte Ltd.
The registered company address is: 152 Beach Road, #21-01/04 Gateway East, Singapore 189721, Singapore

Preface

The goal of this book is to bring together significant international research in technology education by focusing on contemporary PhD theses. An international search was conducted to identify doctoral researchers who have recently completed their theses. The long list was shortened due to availability and a range of personal reasons, and some who joined the project had to later withdraw. This left a group of 16 technology education researchers from 10 countries. Of course, this is not a representative sample of international research in technology education but a snapshot of recent doctoral research from around the world in this area.

An issue with much research which has implications for teaching is ensuring that it is accessible to the people who can use it – teachers. Many doctoral students publish elements of their thesis while they are undertaking their research, often at conferences, and then focus on more substantial journal publications upon completion of their thesis. Nevertheless, the applied message of the research often does not get to the technology teachers who could benefit from the research to inform their practice, and therein lies the intent of this book.

Each author has been asked to write a chapter based on their thesis. Each chapter has a similar structure, with the focus being on what the research means for classroom teachers. While the conceptual underpinnings of each research project are made explicit and the methodology is summarized, the focus of each chapter is an elaboration of the findings in ways that are relevant for practitioners. Each chapter provides a reference for the thesis so a reader can pursue more detail if they would like.

The last section of each chapter provides an indication of the authors thinking about where this research might lead in the future and, thereby, possible areas in which teachers may be interested in conducting further research. It is the editors' wish that this book will help overcome an incongruity of educational research: it is done with the intent of improving practice, but those who do the practice rarely hear about it.

The editors would be delighted to hear from any researchers who have recently completed doctoral-level research in technology education. This is the second volume in this series based on doctoral research, and the third volume is underway. The first volume was *Contemporary Research in Technology Education: Helping Teachers Develop Research Informed Practice*.

Perth, Australia
Brighton, UK

P John Williams
David Barlex

Contents

Contributors

Abbad Almutairi Ministry of Education, Center for Education Polices Researches, Riyadh, Saudi Arabia

Cecilia Axell Department of Social and Welfare Studies, Linköping University, Norrköping, Sweden

David Barlex University of Exeter Associate, Brighton, UK

Scott R. Bartholomew Purdue University, West Lafayette, IN, USA

Eva Björkholm Royal Institute of Technology, Stockholm, Sweden

Sachin Datt Sushant School of Design, Ansal University, Gurgaon, India

Thomas Delahunty University College Cork, Cork, Ireland

Michael E. Grubbs Baltimore County Public Schools, Towson, MD, USA

Eva Hartell KTH Royal Institute of Technology, Stockholm, Sweden

Vicki Knopke Griffith University, Brisbane, Australia

Kieran McGeown St Marys University College, Belfast, Northern Ireland, UK

Sonja Niiranen Tampere University of Technology, Tampere, Finland

Adrian O'Connor Portmarnock School, Dublin, Ireland

Jason Power University of Limerick, Limerick, Ireland

Victor Ruele University of Botswana, Gaborone, Botswana

Mary Southall Western Sydney University, Sydney, Australia

Greg J. Strimel Purdue University, Lafayette, IN, USA

P John Williams Curtin University, Perth, Australia

About the Editors and Contributors

Editors

P John Williams is a professor of education and the director of graduate studies in the School of Education at Curtin University in Perth, Western Australia, where he teaches and supervises research students in technology education. His current research interests include STEM, PCK and electronic assessment of performance. He is the editor of the *Australasian Journal of Technology Education*, advisory editor of the *International Journal of Technology and Design Education* and series editor of the Springer *Contemporary Issues in Technology Education*. He has authored or contributed to over 240 publications and is elected to the ITEEA's Academy of Fellows for prominence in the profession.

David Barlex is an acknowledged leader in design and technology education, curriculum design and curriculum materials development. He taught in comprehensive schools for 15 years before taking university positions in teacher education. He directed the Nuffield Design & Technology Project. David is well known for developing curriculum materials that support pupil learning from a constructivist perspective. He uses this approach to develop young peoples' ability to understand and critique the design decisions made by professional designers and those they make themselves. This informed the Nuffield Design & Technology publications, which have been widely used in the UK and emulated abroad – in Russia, Sweden, Canada, South Africa, Australia and New Zealand.

Contributors

Abbad Almutairi is a researcher from Saudi Arabia; he works as a director of Total Quality Department at the Education Office in Al Mahd Province (Saudi Arabia) . He studied in New Zealand, and he gained three qualifications from three different universities (DipEd, MEd and PhD). All his study in New Zealand was about teaching technology education in New Zealand and Saudi Arabia. His current research interests include STEM, leadership and schooling accreditation. He is a member of the New Zealand Technology Education Association (TENZ) and a member of the founding committee of Technology Education Consultants of New Zealand (TECNZ).

Cecilia Axell is a senior lecturer and researcher in technology education in the Department of Social and Welfare Studies, Linköping University. She holds a PhD in technology education (2015) and has a background as an elementary school teacher in technology, sciences and mathematics. She is also a trained Montessori teacher and taught at comprehensive schools for 18 years before starting her PhD studies and engaging in teacher education at the university. Her research focuses on the pedagogical content of children's fiction which links to the relationships between humans, technology, nature and future perspectives. She is also involved in research projects focusing on technology education in early childhood.

Scott R. Bartholomew is an assistant professor of engineering/technology teacher education at Purdue University. He has taught a variety of subjects ranging foreign language to photography and video production to woodworking. Prior to completing his graduate work, he taught technology and engineering classes at the middle school and university level and coached high school volleyball. He is an active member of ITEEA and has served in several leadership positions coordinating international competitions for technology and engineering education undergraduate students. His current work revolves around Adaptive Comparative Judgment (ACJ) assessment techniques, student design portfolios and technology and engineering teacher preparation.

Eva Björkholm is a former primary and lower secondary school teacher, now working as a lecturer and researcher in technology education in the Department of Learning, KTH Royal Institute of Technology in Stockholm, Sweden. She has been teaching in technology education courses in different teacher training programmes for 20 years and completed her doctoral thesis in 2015. Her research interests include design-based research focusing on primary technology education.

Sachin Datt is an assistant professor at NMIMS School of Design, Mumbai, India. Formerly, he was visiting fellow at Homi Bhabha Centre for Science Education, TIFR, Mumbai. He has published comic books on history of science and technology in native Indian languages while working for NGO called Pratham Education. His

PhD thesis was on using narratives of history of science for teaching science. His graduation was in the discipline of visual communication and undergraduation was in fine arts. His main research work focuses around exploring narratives structures 'of and around' technology and how they can enrich design and technology education.

Thomas Delahunty is currently a member of research faculty at the University of Nebraska-Lincoln where he holds positions as an NSF-funded research fellow working with the Department of Electrical and Computer Engineering and the Department of Psychology. Tom is currently working on a collaborative project looking at the aetiology of spatial cognition in engineering problem-solving through the lens of cognitive neuroscience. This project was designed and successfully funded based on Tom's doctoral work that is discussed in this volume. Tom has earned several international awards for his research including winning his second ASEE Oppenheimer Award in January 2018.

Michael E. Grubbs is a coordinator of career and technology education for Baltimore County Public Schools. Previously, he was the supervisor of manufacturing, technology, and engineering education for Baltimore County Public Schools and a middle school engineering and technology education teacher, graduate research assistant for Virginia Tech and adjunct instructor. He has co-authored the textbook, *Foundations of Engineering & Technology*, coordinated the *Advancing Excellence in Engineering Education* project, was a fellow of the 21st Century Leadership Academy and ITEEA Leader to Watch. Dr. Grubbs current research focus is on P-12 engineering and improving students' design cognition.

Eva Hartell is an experienced elementary STEM teacher and holds a PhD in the area of educational assessment, focusing on teachers' assessment practices. In 2015, her PhD thesis was one of the top ten most read in Sweden. She believes in the need to increase affordance for teachers' assessment practices and conducts school-based research in association with teachers, schools and municipalities. She is currently working in Haninge Municipality and at KTH Royal Institute of Technology in Sweden and is actively involved in the global grassroot teacher lead organism research ED.

Vicki Knopke completed her doctoral thesis at Griffith University, Australia, in December 2015. Her master's thesis focused on technology in primary schools, while the recent study examined senior secondary schools. Vicki worked for Education Queensland for 27 years as principal education officer for school improvement working with administrators to bring about positive change in schools through learning initiatives. Prior to this role, she worked as curriculum development officer implementing the key learning area of technology within schools and as a teacher and senior school syllabus developer for 16 years. Her current research interests are in communities of learners and motivating youth to develop their learning.

Kieran McGeown is presently employed as a senior lecturer/course team leader on the 4-year BEd programme in St. Mary's University College, Belfast, Northern Ireland. He has a keen research interest in matters pertaining to the development of effective interdisciplinary and multidisciplinary pedagogical strategies for STEM education. Kieran has served as an external examiner for secondary PGCE DT programmes in several UK universities and is at present a member of the Northern Ireland Engineering Policy Group. He is also one of two fellowship councillors for the Royal Society for the encouragement of Arts, Manufactures and Commerce (RSA) in Ireland. He has been involved in technology education for the past 25 years.

Sonja Niiranen is a university teacher in technology education. She received her master's degree and her doctoral degree in education in 2006 and in 2016, respectively. She has been working in the Department of Teacher Education at the University of Jyväskylä since 2005 by delivering technical craft and technology education and entrepreneurship education courses for elementary teacher education students. She has also worked short periods in elementary schools as a teacher. Her current research interests are related to learning by doing and creating things with hands in craft and technology education.

Adrian O'Connor is a teacher/practitioner researcher of design and technology education in Portmarnock Community School, Co. Dublin, Ireland. Previously, he was a lecturer of technology education in the School of Education, University of Limerick, Ireland. He is also a working member of the Technology Education Research Group (TERG).

Jason Power is a lecturer in the Faculty of Science and Engineering in the University of Limerick. His research focuses on psychological constructs that influence student performance in STEM learning environments. Jason has previously contributed to large-scale school-based research project in the USA and continues to collaborate with a multi-institutional team comprised of researchers from Northwestern University, University of Cincinnati and the University of California. He has designed and delivered multiple initial teacher education modules dealing with many aspects of technology education and has received national-level awards in recognition of excellence in third-level teaching.

Victor Ruele is a lecturer in the Department of Industrial Design and Technology at the University of Botswana. He has been with the institution since 2006. He teaches design and technology-related courses for pre-service and in-service teachers. This includes the training of D&T teachers from Swaziland and Lesotho. Prior to this, he has worked as a teacher and head of the department in the secondary schools for over 15 years. He contributed towards the development of systems and control content areas of the D&T curriculum. He has been involved in the continuous professional development (CPD) of local teachers in the field of control technology. His research interests include change management, curriculum innovations and technological literacy.

Mary Southall was a design and technology teacher in the UK for over 10 years where she was both head of the department and assistant teacher and gained advanced skills teacher and lead practitioner status. Mary has worked as an education consultant both within the UK and internationally, writing a wide range of resources and supporting teachers in the classroom deliver effective design education. Her doctoral research focused upon assessment of learning in design and technology, and she is currently working at Western Sydney University in Australia as a curriculum advisor and lecturing on both the master of teaching (secondary and primary) programmes.

Greg J. Strimel is an assistant professor of technology leadership and innovation at Purdue University in West Lafayette, Indiana, USA. His efforts are focused on enhancing the appropriately scaffolded teaching of engineering and design at the P-12 level by studying design cognition with respect to student performance and engineering design-based instructional interventions. Previously, he was an instructor of engineering fundamentals at West Virginia University where he also served as the director of K-12 initiatives. Additionally, he has taught secondary-level engineering education and served as an infantryman in the US Army. Strimel has been recognized as a *New Technology & Engineering Teacher of the Year*, *Emerging Leader in Engineering/Technology Education* and *21st Century Fellow* by the International Technology & Engineering Educators Association.

Chapter 1
Introduction

P John Williams

Abstract The goal of this book is to bring together significant international research in technology education by focussing on contemporary PhD theses. An international search was conducted through professional associations and higher education institutions which support postgraduate technology education research, to identify doctoral researchers who have recently completed their theses. Those that were available to develop their thesis into a book chapter constitute the authors in this volume.

The goal of this book is to bring together significant international research in technology education by focussing on contemporary PhD theses. An international search was conducted through professional associations and higher education institutions which support postgraduate technology education research, to identify doctoral researchers who have recently completed their theses. Those that were available to develop their thesis into a book chapter constitute the authors in this volume.

Each author was asked to write a chapter based on their thesis. Each chapter has a similar structure, with the focus being on what the research means for classroom teachers. In order to be confident of the conclusions and recommendations that have been outlined in each chapter, a little about the conduct of the research has been included – what were the research questions, how were they answered and how was the data collected and analysed. However, the focus of each chapter is an elaboration of the findings in ways that are relevant for practitioners. Each chapter provides a reference for the thesis so a reader can pursue more detail if they would like. Each chapter is structured around similar headings.

The focus of this book is on the provision of evidence-informed practice, and we hope this will be useful in two ways. In the implementation of these ideas into their own practice, teachers can be confident that there is a researched rationale underpinning the ideas. Secondly, it can be difficult for teachers to convince school administrators of certain needs in the technology education learning area; these chapters

P J. Williams (✉)
Curtin University, Perth, Australia
e-mail: pjohn.williams@curtin.edu.au

P J. Williams, D. Barlex (eds.), *Explorations in Technology Education Research*, Contemporary Issues in Technology Education,
https://doi.org/10.1007/978-981-13-3010-0_1

provide evidence that may be useful in this context to help ensure that technology education is an effective and essential element of the core curriculum.

The chapters have been grouped into six areas: curriculum content, stories of technology, planning and pedagogy, cognition, girls and technology and information technology.

Curriculum Content

The first group of chapters focus on technology education curriculum – at a macro level in Saudi Arabia, New Zealand and Botswana and at a micro level in terms of student's perceptions in Northern Ireland and their learning in Sweden.

In Chap. 2, Abbad Almutairi sought to benchmark the development and context of technology education in Saudi Arabia through an examination of and comparison with technology education in New Zealand. He explored how technology education is implemented in both contexts as a way to identify the similarities and differences between them. Data were obtained from document analysis, interviews with technology teachers and technology experts, observations of practice and questionnaires, in both countries.

Technology education in Saudi Arabia has been integrated with science subjects, and the concept of technology has not been identified as an independent entity. In New Zealand it is a separate subject called 'technology education' that has a general aim of developing a broad technological literacy that will equip students to participate in society as informed citizens. In Saudi Arabia, on the other hand, greater emphasis was given to rationalizing the teaching of science together with technology, rather than identifying its independence.

Abbad found that New Zealand teachers implemented a variety of teaching methods and that problem-based learning, enquiry learning, discussion, designing and student self-assessment, together with collaborative and individual learning, were the common approaches. The Saudi teachers tended to use more traditional teaching methods such as lectures, discussion and, occasionally, collaborative learning. Technology experts in New Zealand have positive attitudes towards teaching technology education, and they believe it is an essential subject that must be included in the curriculum as a separate subject. In Saudi Arabia, the experts considered that their schools are not ready for technology to be a separate subject.

Victor Ruele examined the development of the curriculum in Botswana in Chap. 3. Although Botswana gained its independence from Britain in 1966, it has until recently followed a British structure in its design and technology curriculum, largely through links with the Cambridge curriculum and examination system. Victor explored the management of the change process from a British model of design and technology curriculum to the Botswana model, in the context of the senior secondary schools. The goals of the study were to identify and evaluate the necessary steps and strategies for managing the change and transition to ensure effective implementation in schools. The conclusion is that in any change process what is needed is a shared vision, creation of a strong and credible coalition, creation of enabling environment, provision of resources and continuous monitoring and professional development of all those involved in the change process.

A multi-method approach was adopted to data collection which included questionnaires, individual and group interviews and document reviews. A number of challenges were identified with the implementation of the new curriculum such as a limited implementation strategy, limited participation by key stakeholders, a weak coalition for change, limited administrative support especially in terms of provision of resources and a limited teacher support system as well as weak reinforcement mechanisms to sustain the change. The major issues which were raised by the teachers concerning the implementation of the curriculum included a 'crowded curriculum'; the need for upskilling or professional development, particularly in electronics, pneumatics and computer-aided design and manufacture (CAD/CAM); and the difficulty of resourcing the equipment needed to implement the new technology content areas. The implementation of an integrated curriculum was one of the major challenges for many teachers.

Eva Björkholm's research in Chap. 4 explored two technology classrooms with students aged 6–8 years in Sweden. In the past in Sweden, technology teachers have largely focussed on technology as an opportunity for students to experience practical work. This means that technology teaching has not, or has only to a small extent, explicitly focussed on what the students are expected to learn. Eva's response to this traditional focus was to develop and describe knowledge in technology in the early school years.

Data were mainly generated through two so-called learning studies conducted in primary schools. Learning study is a classroom-based research approach with a high degree of intervention and collaboration between a group of teachers and the researcher and consists of four cyclically repeated steps, investigating the most powerful way to teach the specific object of learning. The two learning studies were conducted with a total of 49 7–8-year-old pupils and 49 6–7-year-old preschool and grade 1 pupils.

Video recording was chosen for generating data, since the pupils' technical knowledge was understood to comprise skills and understanding as integrated forms of knowledge, that is, without separating thinking and doing. The video recordings were analysed using phenomenographic analysis which resulted in categories describing pupils' different ways of knowing the object of learning. The categories were:

- *Effectiveness for me – fulfilling functions based on own needs and everyday situations. In this category the students related the fitness for purpose only to their own use of the artefacts.*
- *Effectiveness for others – fulfilling functions based on others' needs and situations. In addition to their own use, the students evaluated the artefacts in relation to other people's use.*

- *Construction dependency – how efficiently aspects of the construction help to realize the function. In this category the students also analysed the construction of the technical solution in terms of material, form and components.*
- *Technical efficiency – how efficiently commonly agreed technical solutions realize the function. This category contained the most complex way of understanding, meaning that common technical solutions, for example, the hinge, were identified and compared to other solutions in terms of how efficiently they realize specific functions.*

Based on the categories described above, critical aspects were identified, which are necessary for the pupils to discern in order to be able to evaluate the fitness for purpose of technical solutions:

(a) Functions related to needs of users other than me
(b) Material, form and components and how they contribute to fulfilling the functions. Linking main and secondary functions to the corresponding key components of the construction
(c) Common technical solutions and how they are constructed in terms of interaction between key components

This knowledge can assist in curriculum planning, devising assessment and developing pedagogical approaches.

The final chapter in this section is contextualized in the technology and design curriculum of Northern Ireland. In Chap. 5 Kieran McGeown investigates the perceptions held by pupils, across four Northern Ireland secondary schools, towards routine school-based practical work activities and project work in technology and design.

He examined 44 year 10 students who were involved in building a small-scale model of a hydroelectric turbine, building and programming micro-robots for the purposes of accomplishing specific tasks and designing a device that would enable elderly people to transport their shopping from a car into their home with the minimum of stress. Data was collected through the administration of questionnaires to pupils, the conducting of interviews (industrial experts, students and HoDs) and the observation and recording of practical work and industrial visits. A thematic analysis of the pupils' interviews was carried out, revealing the very positive response of the pupils to participating in practical activities and project work. Some students considered that the practical tasks were too quickly completed or that they were unable to complete them due to a lack of time, and it was a common perception that there was 'too much theory'.

Some of the younger pupils suggested that practical projects were too short, too childish and not challenging enough and were left uncompleted or were irrelevant to their and other's everyday lives.

A number of students observed that, as a result of the industry visit, they were more understanding of the fact that product design is an iterative process, involving modelling at its core, and where failure is regarded both as an integral part of the process and a normal occurrence. The nature of the learning experience varied substantially depending on the industrial context in which it took place. The need for focussed preparation for the industrial visits was apparent, in respect of identifying the intended learning outcomes and sharing these with students. The students did not value either the computer-based research work or the portfolios associated with their projects.

Stories of Technology

The two chapters in this section have the notion of stories in common, though in the different contexts of Sweden and India. It is not difficult to see comparisons with the literature of other cultures, and so the general messages of these chapters are internationally applicable.

Cecilia Axell begins in Chap. 6 with an analysis of a selection of Swedish children's literature. The underlying assumption is a belief that children's literature can be considered a carrier of values and approaches to technology. The issues embedded in the stories include reflections about the nature of technology, discussions about its advantages and disadvantages and the way technology is described in its social and historical context.

Cecilia selected six books, from the early 1900s to the early 2000s, on the basis that they not only contain depictions of technology but also represent issues and problems considered relevant today in the field of technology education. A detailed interpretation of the text and then a comparative analysis resulted in the identification of six different themes:

- Technology as a *metaphor or analogy*
- Technology as *anthropomorphic*
- Technology as *autonomous*
- Technology as a result of a *creative driving force*
- *Masculine technology*
- Technology as *enduring*

A conclusion from the analysis was that children's fiction could be a starting point for creative discussions about the nature of technology and technology's effects on individuals, society and nature in the past, present and future. Cecilia noted that the presence of the technology in the books often conveys an ambiguous message: on the one hand, the technology is fascinating, essential and a result of creativity. On the other hand, technology is something that adversely affects both human relationships and nature. Such analysis could be the basis for engaging discussions in the technology classroom and provide meaningful contexts for learning about the nature of technology. A fictional story can also give an opportunity to introduce specific technological terms and concepts, as well as serving as a starting point for technological activities.

In Chap. 7, Sachin Datt begins with the premise that humans have entertained themselves with stories since the beginning of civilization. In a technology context, he poses the questions: What are the characteristic features of narratives that carry the cultural values of technology? And how can these narratives help in communicating the nature of technology? As the source of data, Sachin analysed the autobiographical narratives of the Wright brothers and Otto and Gustav Lilienthal, the people largely responsible for flying and controlling the first heavier-than-air machines. The analysis resulted in 22 common elements of the stories, for example:

- The lead protagonist or a group of protagonists get introduced to some 'existing' technological artefact that arouses their curiosity.
- The lead protagonist begins to imitate the existing technology in some secret manner.
- The protagonists get involved in the existing technology as a sport.
- The presence of a helping mentor who provides material, technical and emotional support has been seen repeatedly in narratives of flight.
- There is a tension between the protagonists' everyday life which may include family or survival pressure, a regular job and the 'technological sport' that the protagonists are pursuing.
- Science entered the picture of flying machine technology when the Wright Brothers saw flying as more than only a sport.
- Once resolved, the protagonists decide to demonstrate the new possibility to a closed group of people who really matter rather.

Such stories, particularly in the context of design and technology education, can enable students to appreciate the contribution of inventors/innovators of technology as part of their understanding of the nature of technology. Sachin suggests that teachers need to be careful in selecting the right kind of stories for design and technology classes. Not all stories give the correct idea of the essence of events that lead to an innovation. Some stories give genius-like qualities to inventors who come up with ideas in a flash of intuition without giving details of the past inventions that the inventor has referred to. Such a problem is very prevalent in the Indian context where inventions are shown to get created out of some mythical superpower of the genius inventor. The structure of narratives of technology that are presented in this research can help teachers in choosing and selecting narratives that provide the essence of technological innovation events that may be closer to reality.

Planning and Pedagogy

This group of chapters relate to the work of teachers in planning for effective learning, through assessment, focussing on creativity and problem-solving and task planning.

In Chap. 8, Eva Hartell explored the effect of teachers' self-efficacy on assessment, their formal documenting practices, minute-by-minute classroom assessment and their statements and motives relating to criteria for success while assessing

students' e-portfolios. She employed both quantitative data collected from official governmental databases, software-generated statistical data from questionnaires and qualitative methods such as observations, analyses and interviews.

Eva demonstrated that teachers use assessment in different ways. Formative assessment is necessary for bridging teaching and learning in technology, particularly through the inclusion of questions to pose and check pupils' progress during the learning journey. But teachers need support from a number of elements in their educational environment to ensure the progress of their pupils.

Eva concluded that it is important to embed into the teaching process five key strategies for formative assessment: clarifying and sharing learning intentions and criteria for success; engineering effective classroom discussions, questions and learning tasks; providing feedback that progresses learners; activating students as owners of their own learning; and activating students as instructional resources for one another. Teachers must also provide time for students to learn from mistakes and need to be in an environment that supports them, where they can grow their assessment literacy and self-efficacy. The importance of affordance for teachers' assessment practices cannot be emphasized enough.

Mary Southall (Chap. 9) investigated the approaches used by technology teachers to plan how teaching, learning and assessment activities work together to ensure students have the opportunities to demonstrate creativity and the development of creative skills. Her study divided planning into three stages: pre-active, interactive and post active. Seven schools participated, and lesson plans were gathered from years 7, 8 and 9 technology lessons. Other data was collected by email questions and classroom observations.

Mary found that the use of whole-school lesson planning proformas was standardized in the secondary schools she researched but that teachers also used a variety of informal approaches to planning, which tended to be stimulated by a 'thought' often triggered by an inspiration, for example, a product or television programme.

The teachers specified minimal details in regard to formative assessment and the processes for formulating learning, offering far more information on planning for the resources to support the learning intention. 'Teacher talk' was the dominant form of teaching strategy, with two thirds of all activities being described as 'teacher explanation' or 'teacher discussion'. Consequently, the majority of learning demonstrated in the lessons could be classified as either knowledge based or practical skills, producing learning outcomes that were either written, sketched/drawn or produced from a practical activity. Just under half the lesson plans involved either group discussion, group work or paired learning, involving students talking to, or with, other students.

Mary concluded that teachers need to plan the learning environment to support, not dictate, the development of the essential knowledge, skills, understanding and processes or key concepts. Such an approach, which develops knowledge and skills on a 'need-to-know' basis, places an emphasis on teaching students a process that involves identifying how and when knowledge is required and not on the knowledge students may one day need.

Tom Delahunty began Chap. 10 with the assumption that problem-solving skills are a critical element of technology education. His literature review found that traditional research in this area has emphasized processes and heuristic approaches to solving problems while neglecting the early stages of problem-solving such as problem conceptualization. One of the key goals of this research was to investigate if there is a relationship between problem-solving conceptualization and approach. Electroencephalography (EEG) was used to collect and analyse data on problem-solving episodes through the use of headsets. This data was enhanced by interviews and a Likert scale rating of the mental effort required to solve each problem. The subjects were 12 third-year undergraduate initial technology teacher education students.

Tom developed three different categories of problem-solving tasks: mathematical, visuospatial and a dual approach. The EEG indicated that once a participant conceptualizes a problem, at least from a cognitive perspective, their approach is very much dependent on and fixed to that conceptual frame; this finding was reinforced by the other data.

A lack of metacognitive regulation and understanding was evident in the participants' approaches to solving the problems. It is concerning that these students display little of this attribute, as one of the hallmark attributes of expert problem solvers is that they are adaptive. Reliance on past experience and in particular experience informed by the participants' past schooling was a common trend.

It seems that these conceptual rigidities that have been (at least somewhat) cultivated in school are pervasive and effect students much further into their academic career. The majority of problems that are encountered by students are convergent or well defined in nature, which are typically suited to a particular approach and have a single correct solution. This focus does not provide students with the breadth of experiences necessary to develop skills in creative engagement with ill-defined problems.

In order to develop more flexible conceptualization skills in students, it is necessary to provide an environment where problem-solving approaches can be explored, communicated, critiqued and evolved in a collaborative and constructive manner. Perhaps the most simple and pragmatic way to do this is to ask them to introspect on the process. What was apparent to Tom is that the overemphasis on specific problem-solving approaches and an over-adherence to a specific task design can limit students' opportunity and propensity to develop flexible and adaptive problem-solving skills.

Jason Power examined the impact of a person's belief about their own capabilities and how this influences their performance in Chapt. 11. His main question was how does difficulty influence self-efficacy? This led to thinking about how reward influences a person's behaviour, and if increased reward could result in increased engagement, could it be a way to coax students to be more engaged? Jason then wondered whether it would be possible to cancel out the negative effects of increased difficulty with the positive effects of increased reward.

To try to answer these questions, Jason had 240 students (av. age, 12.4 years old) play four rounds of Pac-Man during a school week on their personal USB version

of the game. At the start of the week, each participant would play a standard version of the game (to gauge initial ability level) and again at the end of the week (to gauge improvement). What the participants did not know is that they had been given one of three versions of the practice task with a subtle manipulation of either difficulty or reward based on the round.

Those who practised with the highest difficulty consistently performed worse than the group that practised using the easy version of the task, and lower difficulty leads to increased engagement. But it also led to higher levels of mastery experience (claimed to be the most powerful source of self-efficacy), vicarious experience and physiological state. Jason's conclusion was that self-efficacy could be manipulated by difficulty.

He does point out however that it is vital that more complex concepts are scaffolded in such a way as to provide opportunity for the student to experience mastery, which is a potent source of self-efficacy. Descriptive feedback surrounding how the individual has improved relative to previous performance can also have a profoundly positive impact on a student's self-efficacy.

Cognition

Greg Strimel and Michael Grubbs are members of a research group in the USA which is conducting research into how students think. Their doctoral research presented in these chapters is foundational to further research that is taking place in this generally under-researched area of technology education.

Greg began Chap. 12 with the belief that it is critical to understand the ways in which students mentally process engineering design tasks in an effort to provide effective instruction, establish the adequate scaffolding of design experiences and enact interventions to enhance student design abilities. In this research eight high school engineering students were tasked to verbalize their thoughts while completing a design task. Participants were equipped with point-of-view cameras that gathered video of their non-verbal cues and their verbalized thought processes. These videos and the participants' design artefacts (design journals, prototypes and teacher-scored project rubrics) enabled triangulation of data while conducting the data analyses. This enabled the researcher to compare each participant's cognitive strategies with the effectiveness of their final solution. The videos were segmented into individual cognitive tasks and coded.

The analysis revealed that the top 3 most commonly employed mental processes were *model/prototype constructing*, *analysing* and *managing*. The least used mental processes were *experimenting* (0.7%), *computing* (0.08%), *questioning/hypothesizing* (1.9%), *defining problems* (2.0%), *interpreting data* (2.1%), *predicting* (2.3%) and *measuring* (2.6%).

The participants who devoted more time employing the scientific-related mental processes (*testing*, *experimenting*, *observing* and *interpreting data*) produced better performing solutions. The top-performing participants enacted more iterations of

testing their solutions, making observations, interpreting the outcome data and then experimenting with design changes to improve their solutions than the poor-performing participants. The participants who created the best-performing solutions noted in their design journals that they related the challenge to some other experiences they have had.

As a result of this research, Greg recommended that teachers should teach effective project management and process planning techniques when implementing engineering design-based lessons and should emphasize the application of developmentally appropriate mathematical practices, computational modelling techniques and quantitative data analyses. Students should be provided with specific authentic tools, materials and design experiences that can be drawn upon to solve other problems in the future.

Michael Grubbs outlined that the purpose of his Chap. 13 is to present strategies and heuristics on how to provide feedback through assessments (preformative, formative and summative) and how design challenges can be used to situate students within a context that develops their cognitive abilities. He proposed the question: How do students distribute their cognitive efforts as they solve problems and synthesize, analyse and evaluate during engineering design tasks? He examined 40 students, 10 pre-engineering dyads that had taken pre-engineering courses and 10 dyads that had not and were tasked with solving a design-no-make problem to design an assistive technology device within 45 min. Data was captured verbatim from the verbalized thoughts of each student.

Findings from the research indicated that the instructional and pedagogical approaches applied to the engineering students do not appear to prepare them to any greater degree than students who do not take pre-engineering-related coursework, for engaging in the design problem. As a result, Michael concluded that educators should purposefully present strategies and heuristics for students, improve feedback through assessments (preformative, formative and summative) and more deliberately consider how design challenges can be used to situate students within a context that intentionally and appropriately develops their cognitive abilities. The implication is that technology and engineering education should provide students with explicit instruction and strategies on design issues.

Girls and Technology

While the gender balance of students in technology education has improved over time, there remains an imbalance, and more research is needed in order to continue to encourage girls to be involved in technology education.

Sonja Niiranen's problem in her Chap. 14 was how to increase girls' interest and motivation in technology education by highlighting some practical suggestions for developing technology education, thereby increasing girls' access to and interest in technology education at the basic education level and ultimately increasing the number of women who enter technology-oriented fields. Sonja collected data using a questionnaire study for students in the fifth and sixth grade (n = 281, 144 girls and

137 boys), 12 female technical craft and technology education teachers and 12 female engineering students, exploring the main factors that had influenced women's decisions to study and enter a career in a technology-oriented field.

A deeply gendered history of technology studies and the discourses relating to gender in technological activity have labelled technology as a masculine and male-dominated arena; however all children should have the opportunity to develop their technological literacy. Teachers play a key role in dismantling gendered practices and renewing the image of technology education because they are well placed to alter students' perceptions. Sonja found that there are differences in girls' and boys' motivations concerning *the contents of technology education* and how important students perceive the subject to be. Boys and girls were equally happy about creating their own ideas. It was evident that *social interaction* between the teacher and the students in class was an important factor, especially for girls.

Women who were studying at the university level and later working in technology-oriented fields did have a high level of competence related to the field they had chosen to study or work in. They had received significant intellectual capital regarding technology or engineering in their childhood, which was influential in their occupational choices.

In Chap. 15 Vicki Knopke also recognized that females are significantly under-represented in technology education classrooms in senior secondary schools, and this has a feedforward effect on females entering subjects such as engineering at the tertiary education level. The aim of Vicki's study was to examine the factors that encouraged and facilitated female students to participate and engage in technology education.

She examined year 11 female technology education students and looked for shared patterns of behaviour and group interactions and similarities, which occurred across the three school sites (each one a case study) over a 6–8-week time period. Data was collected through interviews (students, teachers and HoDs), observation data, audio recordings and photographs.

One recommendation from the research was the need to build pedagogical ecologies for technology education, based on an awareness of learning styles and values that are unique to females' ways of learning. A female-oriented pedagogy that meets mixed learning styles with structure but provides individuals with independence to problem-solve and discuss issues appears to be an undervalued aspect of technology education planning.

For this subject to be socially valued, there is a need to incorporate the values and language of diverse participants, and teachers need to re-examine some of their practices in order to entice female participants in technology. It was a personal motivation more than an employment motivation that motivated female students to participate in technology education. Peer support (not limited to same gender) and shared group goals enabled females to excel in some areas of technology.

This research study has shown that strategies to promote female participation involve long-term planning, short-term immediate support and constructionist considerations. Teachers need to be mindful that there is no such thing as a 'gender-neutral curriculum'. The making of jewellery boxes instead of tool boxes does not address this issue.

Information Technology

The incorporation of information technologies into teaching continues to be an issue in education that requires research in order to indicate the effectiveness of technologies and how they can be incorporated into pedagogical strategies. As technology education is imbued with a broad range of technologies, it is a special case in terms of the incorporation of information technologies. The first of the final two chapters in this book focusses on relating technology to the pedagogy of discussion, and the final chapter examines the effectiveness of mobile technologies.

In Chap. 16, Adrian O'Connor examines the design and implementation of a conceptual model which uses learning protocols to support communication between teachers and pupils. He recognizes that with the latest advancements in ICT and cloud-based software, education is no longer confined to a 'classroom' setting, as new configurations for the delivery of education and instruction are possible in any conceivable setting. Adrian analysed the communication of seven classes over 6 weeks, during which time student used a virtual learning environment (VLE) to comment; attach various text, image, audio or video files; and share links.

The data for analysis was the record of the students working iteratively using the VLE, tagging posts and comments with learning protocols, collaborating with others and uploading data files as evidence. This data allowed this research to track, manage and record the complex and fundamentally non-linear nature of teachers' and pupils' activity during the project.

The teachers and pupils responded positively to the VLE; it stimulated their interest, motivation and engagement in the regular day-to-day school activity. It also helped to extend learning and enhance practice as it promoted more scope for periods of critical thinking and reflection. The data provided evidence of an iterative and interactive process of learning. Pupils had more time to both ask and answer questions on the issues they found to be personally meaningful and educationally worthwhile, which ultimately guided pupils in the development of more robust ideas and solutions. It provided opportunities for pupils to evidence their learning and to make their learning more visible, which supported teachers and pupils alike in planning their intervention strategies and basing their decisions on evidence.

In the final chapter, Chap. 17, Scott R. Bartholomew presents the findings from research with middle school students working in small groups on an engineering design challenge. The research centred on the relationship of access to mobile devices and student self-directed learning as well as student achievement. Seven hundred middle school students at five schools (18 classes, 6 teachers) worked in small groups of 2–3 during an open-ended engineering design activity. The study took place over five class periods. Two teachers used paper portfolios with their classes, while four teachers used iPads to complete identical electronic portfolios. Data was collected using pre-study questionnaires and images of student design products. Each paper portfolio was also 'digitized' into electronic portfolios.

The research found that two significant predictors of student self-directed learning scores were average mobile device skill level and computer access and use at

school. Students with a higher skill level in using mobile devices were more self-directed in their learning, and students were more self-directed after the study than prior to the study taking place. This hints that participation in open-ended problems may lead to an increase in student self-directed learning.

The students who used a paper portfolio received significantly better ranks than their peers using the electronic portfolio, and students with access to mobile devices ranked better than their peers.

The findings, teacher comments and researcher observations all appeared to indicate that mobile devices may not be an influential factor in student success or self-directedness in learning. Scott found that access to mobile devices did *not* have a significant impact on student self-directed learning, and access to computers at school was *negatively* correlated with student self-directed learning.

To best assist students in becoming more self-directed in their learning, teachers can not only equip their classrooms with technology and other tools; they should also teach specific skills such as working in groups, solving open-ended problems and technology tool use. There is a need for direct instruction of students regarding how, where and when to use their mobile devices. Teachers should work to explicitly teach students ways that mobile devices could be leveraged to perform tasks other than simply finding facts.

school. Students with a higher skill level in using mobile devices were more self-directed in their learning, and students were more self-directed after the study than prior to the study taking place. This infers that participation in open-ended problems may lead to an increase in student self-directed learning.

The students who used a paper portfolio received significantly better ranks than their peers using the electronic portfolio, and students with access to mobile devices ranked better than their peers.

The findings, teacher comments and researcher observations all appeared to indicate that mobile devices may not be an influential factor in student success or self-directedness in learning. Scott found that access to mobile devices did not have a significant impact on student self-directed learning, and access to computers at school was not even correlated with student self-directed learning.

To best assist students in becoming more self-directed in their learning, teachers can not only equip their classrooms with technology and other tools; they should also teach specific skills such as working in groups, solving open-ended problems and technology tools. There is a need for direct instruction of students regarding [illegible]

Part I
Curriculum Content

Chapter 2
Technology Education in Saudi Arabia in Comparison with New Zealand: Lessons for Technology Education Teachers

Abbad Almutairi

Abstract This research sought to obtain a deeper insight into technology education implementation in New Zealand and Saudi Arabia in order to explore how this subject is implemented in both contexts that leads to the identification of the similarities and differences between them. A qualitative approach was used in the main but supported by a qualitative approach in the Saudi context. Data were obtained from document analysis, interviews with technology teachers and technology experts, observations of practice, and questionnaires. The findings indicate that there were significant differences in implementing the two subjects in both contexts, particularly in terms of practice. Limited similarities between them were also found.

This study aims to provide teachers with new data gained from research about different aspects of technology education, namely, curriculum planning, teaching methods, assessment for learning, and student engagement, that might help teachers to improve teaching and learning in the technology education classroom.

The Questions I Asked and Why I Think They Are Important

The purpose of this study was to overcome the problem resulting from the absence of a clear concept and content for technology education in Saudi primary education. This was achieved by analyzing the situation of technology education in Saudi Arabia and New Zealand, exploring the similarities and differences between the two systems. To address the general purpose of this study, I posed the following principal question and sub-questions:

How can the experience of New Zealand's implementation of technology education inform technology education in the Saudi curriculum?

A. Almutairi (✉)
Ministry of Education, Center for Education Polices Researches, Riyadh, Saudi Arabia

P J. Williams, D. Barlex (eds.), *Explorations in Technology Education Research*, Contemporary Issues in Technology Education,
https://doi.org/10.1007/978-981-13-3010-0_2

1. How is technology education implemented in the New Zealand primary curriculum?
2. What is the current situation of technology education in Saudi primary education?
3. What are the similarities and differences in technology education in both contexts?
4. What could improve the teaching of technology education in Saudi primary education?

These questions are important because they might help to answer teachers' inquiries about technology education in New Zealand and Saudi Arabia and also to help them use this data to develop their own knowledge to improve technology education practice in their schools. Thus, the study included the following issues that teachers might need to be aware of:

1. To explore how technology education is implemented in the New Zealand primary curriculum. Technology education has been implemented internationally in different ways, for example, it may be a separate subject or embedded within science subjects. Thus, exploring this aspect will provide us in Saudi Arabia with information that we need to decide which direction might be suitable to implement this subject in the national curricula.
2. To explore the current situation of technology education in the Saudi primary curriculum. This subject has not previously been researched, so this study has found what it looks like and how it can be improved to provide students with practical skills via technology education, which has become an important topic in this technological era.
3. To explore the similarities and differences between technology education in the New Zealand primary curriculum and science and technology in the Saudi primary curriculum. Determining the similarities and the differences provides a useful review of technology education in Saudi and suggests a model that fits with the educational environment.
4. To inform the Saudi context of the New Zealand streams of teaching technology based on the New Zealand experience. New Zealand teachers have been teaching this subject for a considerable time, and this has allowed New Zealand to share its experiences that can be used as an example for those who want to review the curricula in their own educational environment.
5. To introduce some recommendations to improve technology education in the Saudi context.

This study is important because it analyzed the issue of technology education in Saudi Arabia in light of what has already been done in New Zealand to inform its own curriculum development. The development of technology education differs from one context to another, so this study was conducted to help develop this subject in Saudi. Saudi, with its new "Saudi Vision 2030," aims to develop teaching and learning through improving the current curricula as a response to the economic challenges of the twenty-first century.

How I Tried to Answer the Questions

To answer the research questions, I used a qualitative approach (comparative case study method) to describe and to compare the current situation of technology education in primary schools in New Zealand and Saudi Arabia, supported by a quantitative approach in the Saudi context. The mixed method approach was used to produce a study that includes deep and broad knowledge, to "capitalize on the strengths of two approaches, and to compensate for the weaknesses of each approach" (Punch 2005, p. 240), and to provide broadly reliable and consistent data (Bryman 1988). Greater weight was given to the qualitative approach throughout the data collection (Morgan 1998).

In terms of the qualitative approach, I collected data using three main methods:

1. Documentary analyses

   Two formal documents were used in New Zealand as major documentary data: (1) the National Education Guidelines (NEGs), which reflect educational policy in New Zealand (Ministry of Education 2008), and (2) technology education in the New Zealand curriculum (Ministry of Education 2007a). In the Saudi context, two formal documents were also used as major documentary data: (1) the educational policy in the Kingdom of Saudi Arabia and (2) primary science and technology textbooks.

2. Interview

   I used semi-structured interviews (face-to-face with open-ended questions). Two technology teachers and two technology experts from New Zealand were interviewed, along with two science and technology teachers and two science and technology experts from Saudi Arabia. The study also used mixed methods: as Guest et al. (2006) indicated, when mixed qualitative methods are used, a small number of interviewees are adequate.

   The questions were developed to ask technology teachers and technology experts about their perceptions regarding the teaching of technology education. The participants were asked the same questions in both contexts.

   The data were analyzed manually through Key Word in Context (KWIC), as suggested by Ryan and Bernard (2003), to identify, verify, confirm, and qualify the themes and categories that emerged from the interviews. I have identified key words and then systematically searched the corpus of text to find all instances of the word or phrase. Each time I find a word, I make a copy of it and its immediate context. Themes get identified by physically sorting the examples into piles of similar meaning.

3. Classroom observation

   The third source of data was classroom *observation.* I selected one primary school in New Zealand and one primary school in Saudi Arabia to observe how technology education was taught in the classrooms. This enabled me to have a better understanding of the context within which technology education occurs. I developed a classroom observation protocol that included the following four main components: general information, learning environment, process of teach-

Table 2.1 The sample of Saudi participants

Participants	Location of data source	Number of participants
Science teachers	Primary schools in Almahd Province	22
Tutors	Technical and vocational institute in Jeddah	17
	Technology college in Makkah	10
	Technology college in Jeddah	11
Technology advisors	Technology Council of Makkah Region	22
Total		82

ing and learning, and indicators of progression. During these observations, I took field notes and interacted with teachers and students during the actual lessons in technology education. All observation sessions were digitally recorded, which helped in the later analysis of the data.

In terms of quantitative inquiry, I developed a questionnaire that was designed based on my own experience and the literature review (Al-Ghamdi 2012; Al-Ghamdi and Al-Salouli 2013; Fath-Allah 2006; Lebeaume 2011; Jones and Carr 1992; Gravemeijer and Baartman 2011; de Vries 2012). Table 2.1 shows the questionnaire's participants in the Saudi context.

What I Found

This section discusses the results gained from the documentary analysis, classroom observations, and interviews.

Findings from the Documentary Analysis

The document analysis found that the educational policies of New Zealand and Saudi Arabia indicated that technology education is important because it provides students with the essential skills they need to understand technology. The New Zealand educational policy has stated the importance of prioritizing the teaching of science and technology. Interestingly, the analysis of the educational policy showed that there was more emphasis on the importance of including and teaching technology in the Saudi educational policy than in its New Zealand counterpart, but the findings of this study found that there was a significant gap between theory and practice in the Saudi context.

The documentary analysis showed that technology education in New Zealand is a separate subject called "Technology Education" that has its own statement, represented in a document called "Technology Curriculum Support" developed by the Ministry of Education (2007b) based on a Technology Education Document

(Ministry of Education 1995). This statement supports teachers and students in the implementation of technology education and is one of the main resources for technology education.

In contrast, technology education in Saudi Arabia has been integrated with science subjects, and the concept of technology has not been separately identified. This integration is a part of the policy of education in Saudi Arabia, which considers science and technology to be integrated subjects. This notion is not explicitly declared in the policy but can be derived from the policy's wording. Although science and technology have very different purposes and methods in education, these differences have not been identified in the policy. This has led to an ambiguous relationship between the two concepts in the Saudi national curriculum.

The rationale for teaching this subject was articulated in the official documents of the New Zealand curriculum and in Saudi science textbooks. By comparing both rationales, I found that the general aim of teaching technology education was clearly identified in the New Zealand curriculum: it aims to develop a broad technological literacy that will equip students to participate in society as informed citizens. In Saudi Arabia, on the other hand, greater emphasis was given to rationalizing the teaching of science together with technology.

The findings indicate that the objectives of technology education in the New Zealand curriculum were clearly identified for each year, from Year 1 to Year 13. An analysis of science textbooks from primary education in Saudi, however, confirms that the objectives did not show which ones are related to science or to technology: this is a result of the integration of both subjects.

The curriculum developers did not identify which skills represent science and which represent technology in order to help teachers select whether to teach them separately or to integrate them in the classroom. In addition, the curriculum developers declared that the philosophy of the science textbooks was to attempt to link science and technology with other disciplines such as mathematics.

One significant difference is related to the structure of technology education in the curricula of the two countries. In New Zealand, the subject is based on three strands that form the technology education concept: the technological practice strand, the technological knowledge strand, and the nature of technology strand. Each of these strands has specific components that reflect them. Also, there are many areas of learning of technology education, such as control, food, ICT, and structural and biotechnology. On the other hand, the subject has no specific structure in Saudi Arabia, but I was able to derive the structure from the introductions in the textbooks, which reflect the philosophy of science and technology. There are two main aspects – scientific knowledge and practical skills – that can be considered as the structure of science and technology. However, scientific knowledge is different from technological knowledge, and, according to Compton (2004), the purpose of technological knowledge is not to make claims to "truth" in the same manner as scientific knowledge does; instead, technological knowledge attempts to understand the process of technological development. The current work indicates that there is no specified content of technology education in New Zealand, but a school's technology department can decide what topics will be taught and uses the New

Zealand education curriculum as a guide to teach technology. In Saudi Arabia, each science and technology textbook includes units and topics that must be taught in the classroom.

Findings from Classroom Observations

The study's findings also indicated that the teachers implemented a variety of teaching methods and that problem-based learning, enquiry learning, discussion, designing, and student self-assessment, together with collaborative and individual learning, were key strategies implemented for technology education in New Zealand classrooms. The Saudi teachers used more traditional teaching methods such as lectures, discussion, and, occasionally, collaborative learning. I argue that these limited teaching methods led to students being taught theory without engaging them in practical activities that would help them to understand the nature of technology.

The way of teaching science and technology in the Saudi schools appears to show evidence of a gap between theory and practice in terms of the science and technology curriculum. The practical activities were indicated in the objectives of science, but the teachers were not provided with guidelines, such as is the case in New Zealand, that help them to practically teach these activities, including teaching steps, suggested teaching methods, and ways of integrating technology with science in the classroom.

The students in both contexts were taught authentic topics, which reflect one of the important aspects of technology education (Turnbull 2002). Turnbull argued that, by involving students in activities that are authentic to technological practice or real-world technology, teachers are able to provide interesting and relevant learning for students. For example, in the observed Year 2 classroom in New Zealand, students were engaged in authentic learning "designing a gecko enclosure" by encouraging them to practice a task that was relevant to their culture and life. This topic was a good example of providing students with authentic learning. They saw the gecko, they participated in setting the goal of the lesson, and they were involved in choosing the materials that would help them solve the problem that the gecko would face, that is, building a home. In the observed Year 5 classroom, the topic was about "alternative energy" that reflected the authentic learning and was relevant to the students' lives. The unit aimed to help them understand how to use energy wisely and to be a good example of conserving energy in their communities (homes, schools, and public places).

Findings from Interviews

Teachers in New Zealand believe that design, and the hands-on concept, make technology education enjoyable because students have the opportunity to touch things and to think about them. Also, the New Zealand teachers were aware of the relationship between science and technology, which primary teachers can integrate in their teaching if they like. The teachers in Saudi Arabia also recognized this relationship. For example, one of the participants explained that environmental issues are "science," while technology focuses more on making machines and devices. However, the Saudi teachers' understanding of this relationship does not mean that they teach science and technology together effectively, and this is what the present study found.

The availability of resources (especially curriculum guidelines and statements) in the New Zealand context helps teachers to enjoy teaching technology and to teach it as it should be taught.

In New Zealand, the teachers indicated that technology education is clearly defined, and they recognize the development of this subject. In the Saudi curriculum, the subject has been integrated with science. It was clear from the learning outcomes I noticed in the classrooms in Saudi Arabia that the Saudi students do not learn much about technology.

The data shows that technology teachers in Saudi Arabia face certain difficulties that preclude them from teaching technology as they would like. Crowded curricula, school timetables, and a lack of professional development were the main problems in this regard. Some of the difficulties that teachers in New Zealand face included timetables, crowded curricula, a lack of resources (materials used to make products) and budgets, and a lack of professional development.

In general, technology experts in New Zealand have positive attitudes toward teaching technology education; they believe it is an essential subject that must be included in the curriculum as a separate subject. These experts consider the unique knowledge involved in technology education and the different philosophy of this subject as the main reasons for this subject being a separate subject in the curricula. In Saudi Arabia, the experts opined that their schools are not ready for technology to be a separate subject.

The experts in both countries had varying opinions about the issue of integrating technology education with other subjects. Andy from New Zealand and Tariq from Saudi Arabia were opposed to the idea of integration, whereas Kate from New Zealand and Ibrahim from Saudi Arabia support integration.

How This Might Be Used to Improve Teaching and Learning

The results of this study might contribute to improve teaching and learning of technology education by helping teachers to consider these results as a source to develop their knowledge about the current issues of technology and to understand the relationship between science and technology. That will lead to enhanced teaching and learning in technology education classroom in terms of the following.

Curriculum Planning

Planning for technology education curriculum is the first step that teachers should consider to ensure that the implementation of this subject reflects the congruence between the goals and practices. At this crucial step, as educators we should consider the following two key questions:

What method should we choose to teach this subject?
What are the justifications for choosing this method?

There are a range of methods for implementing this subject: teaching technology education as an independent topic, integrating it with science, or teaching it across the whole curricula. This study reveals that this is a significant issue in Saudi Arabia, because it has not been defined in the curriculum, and the Ministry has not discussed this issue, neither has it provided teachers with any information about this issue. The justification for choosing to integrate science and technology can be seen in the philosophy and framework of this subject in both the international curriculum and literature of technology education that can be used as a guide to implement this subject. Such a guide should include the following:

1. Definition of science and technology education
2. What is the model and why was this model chosen?
3. General goal of teaching science and technology education
4. Structure of the subject
5. The content and specific objectives of teaching each unit
6. The pedagogy
7. Resources and materials
8. Assessment
9. Professional development

These aspects must be clearly defined in order to clarify what we deem important in technology education.

Identifying the Relationship Between Science and Technology Education

There has been an issue of the relationship between science and technology among educators especially the decision-makers who refer to science and technology as a single entity with the result that science becomes privileged over technology. For example, science has been given more sessions, more activities, and more resources. This has led to a concept of technology as applied science, and it has created a great deal of debate with regard to the relationship between science and technology in the curricula (Kipperman 2006). According to Gravemeijer and Baartman (2011), this debate is caused by the absence of a clear agreement among scientists and technologists on the definitions of science or technology or the relationship between science and technology. Thus, it has been quite common for people "to talk about 'science and technology' as if it were one thing with a double-barrelled name" (Sparkes 1993, p. 25). Cajas and Gallagher (2011) referred to a cluster of articles published in the *Journal of Research in Science Teaching* (2001, 38(7)) that analyzed the relationship between science and technology. The summary that emerges from these articles is that there is a complex relationship between science and technology, and "such complexity should be reflected in the school curriculum" (Cajas and Gallagher 2011, p.713). Cajas and Gallagher (2011) called for a reevaluation and restudy of this relationship. De Vries (2001) explained that the study of the complex relationship between science and technology can be pursued by tracking the history of industrial research laboratories and provides a good opportunity for understanding why and where such complexity exists. I believe that it is essential to reveal these points of view about this issue, to examine them in order to determine the relationship between science and technology, and to deliver this message to everyone who is interested in this field – especially science and technology teachers – in order to improve their understanding of this contentious issue. Thus, in this chapter, I developed a framework in Fig. 2.1 that represents the relationship between science and technology and was derived from the review of literature related to science and technology education.

From the framework, technology and science can be seen as separate disciplines that have overlapping and interacting elements. As such, they require different curricula, although an understanding of the relationships should improve the teaching of each. Almutairi et al.'s (2014) model of the relationship between the two (see Fig. 2.1) shows this separation but also indicates points of interaction. The two major divisions (the first for key aspects of science, the second for technology) represent the argument presented in this chapter that the two fields need to be considered as distinct but are linked as represented by the arrows. An understanding of these components and the relationships between them should provide teachers with a framework to support the development of curriculum and teaching plans. This section provides an overview of the components of the model and further details about the proposed points of interaction.

The top arrow between the two fields represents an important feature of the relationship between technology and science, namely, that scientific knowledge sometimes develops from improvements in technology. Science teachers who understand the limits of testing scientific theories will be better able to inform their students about such theories. Although many theories have been developed from testable hypotheses, some major scientific theories were sound and justifiable before technological advancements provided the tools for formal testing.

Similarly, a technological need may lead to scientific advancements. Areas of scientific knowledge may have some gaps that appear odd because the needs at the time were elsewhere. Present-day improvements in scientific understanding of how materials interact have much to do with the need to develop structures that can withstand assault; and modern environmental science is as much technological advancement as pure theoretical testing. However, theory and tools will both clearly

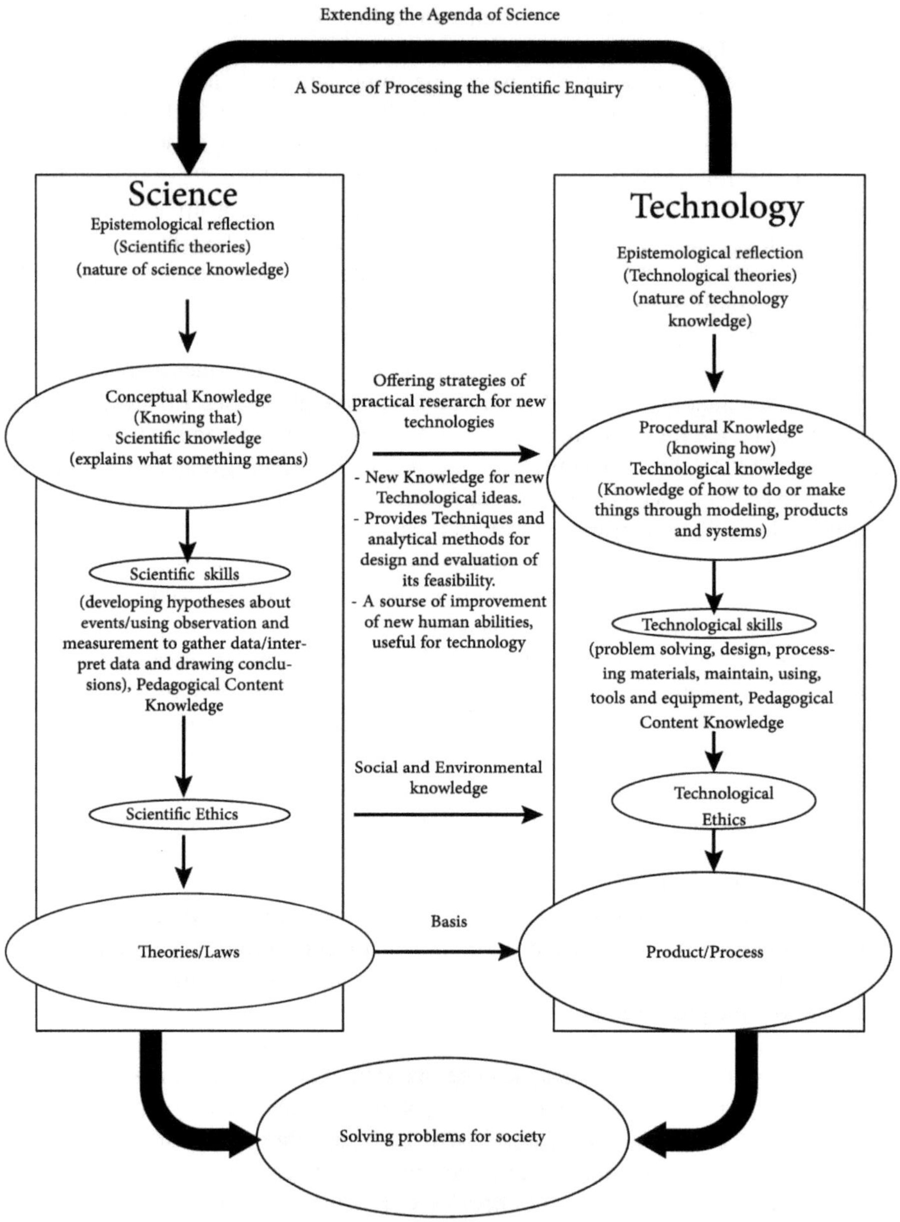

Fig. 2.1 Almutairi's model of the relationship between science and technology. (Source: Almutairi, A., Everatt, J., Snape, P., & Turnbull, W. (2014). Exploring the relationship between science and technology in the curriculum. *Australasian Journal of Technology Education, 1*, 49–63)

be bound by the needs of society. Hence, the arrows at the bottom of Fig. 2.1 serve as a reminder that both scientific theory and technological advancement in such areas as environmental science will be limited, or motivated, by societal views about their importance and the success of the solutions developed. Again, an understanding of these interrelationships will provide teachers of both science and of technology with tools to better understand their student's needs.

Technology is also advanced – and limited – by scientific knowledge, as represented by the smaller arrows in the middle of Fig. 2.1. Teachers of technology would be limited if they did not understand some elements of the theories that led to the development of a tool/product. However, even the simple process of production is often bound by what we would see as scientific methods. It is rare for production techniques to focus on random trial and error. Production typically follows the principles by which scientific methods are bound. The methods for testing a theory are related to the methods that would be used to develop and test a product. Knowledge, theories, and even skills developed in science will form a basis upon which to develop technological advancements. Again, teachers who have an understanding of these relationships should be better able to impart understanding to their students, along with the ability to seek further knowledge.

Any developed model can be criticized, for instance, in this case, one could argue that the technology in the present model should also have conceptual knowledge that is not limited to science. While I agree with this argument, the model indicates that *conceptual* knowledge is significantly connected with science and aims to interpret the events based on theory, while *procedural* knowledge is a characteristic associated with technology that aims to use theory and laws to create technological products.

Fundamental Components of Teaching and Learning in Technology Classroom

When I observed teaching and learning in technology education classrooms to explore how this subject is taught in New Zealand and Saudi Arabia, one of the criteria is related to indicators of progression of teaching technology.

The observation took place in a Year 6 classroom in New Zealand as an example of what teachers need to think about when they want to plan to teach any unit of technology education. The topic was about robotics using sensors, and the focus was on control (Level 3, according to the New Zealand curriculum, in which there are 8 levels of achievement objectives relating to years at school of every subject).

The following seven components were extracted from the observation and are important to enhance the process of teaching and learning in technology education classrooms:

1. Relation of materials to lesson objectives

   Building a moving robot requires a variety of materials that help students to complete this project. The teacher discussed with the students what materials they need and talked about Lego Nxt that included different materials to run the robot. According to Technology Online which is supported by the Ministry of Education (TKI) as an indicator of progression, the teacher could provide resources including a range of appropriate materials, components, software, hardware, equipment, and/or tools for students to select from and guide students to select those that will be suitable for the desired outcome. Offering these materials and encouraging students to select what they think is suitable for their products will improve teaching and learning, and that will reflect on the students' performance and understanding of technological issues.
2. Identifying lesson objectives

   The teacher described the expected learning outcomes that reflect the achievement objectives. For example, he asked his students what they intended to do, why, and how.

   Students could achieve five learning outcomes after finishing the unit: (1) build a moving robot; (2) program a robot to move in a very precise manner; (3) program light, radar, and other sensors; (4) reflect on performances, revise programs, and show analytical skills and perseverance to complete the challenges (ten challenges); and (5) design and build a robot to solve a relevant problem at their school. Six students out of 26 were able to complete up to 9 challenges, while the majority did not exceed the seventh challenge. Students were able to describe the physical and functional nature of the outcome they were going to produce and explain how the outcome would have the ability to address the need or opportunity. This related to what students should achieve based on the progression indicators. Giving students the opportunity to identify lesson objectives enhances teaching and learning through allowing them to establish a conceptual statement that justifies the nature of the outcome and why such an outcome should be developed.
3. Students' learning style

   Discussion and individual and collaborative learning were evident in this session. The teacher introduced the topic by discussing what the robot is and what it is used for. He then encouraged students to learn individually how they could construct the robot and then to work together to help each other to build the robot to make it ready to run. The teacher supported different learning styles that the students were familiar with. Applying more than one learning style enhances teaching and learning through using different teaching methods and helps to alleviate boredom and make the classroom more fun.
4. Students' prior knowledge

   The teacher considered the students' prior knowledge about robots in this session. He started the introductory lesson by asking "Who did the robot last year?" Then he asked them about the basic things they had learned about robots, such as the USB and the brick including its parts (codes, motors, cables, and the sensors). The teacher divided the students into groups and placed in every group at least two students who had information about the robot, to help their peers. Considering students' prior knowledge improves teaching and learning because it helps the teachers to modify the content of the unit based on the students' particular background, and it also helps teachers to recognize those students who have previous knowledge to help their peers and others who might need support during their learning process.

5. Using technological terms in the classroom

   Using technological terms is one of the purposes of teaching technology education so students become technologically literate. Teachers should think about appropriate technological terminology which they should use frequently with their students. I observed that both the teacher and the students used several technological terms in this session, including design, model, properties of materials, problem solving, systems, programming, challenges, and thinking.

6. Engaging students in an authentic task

   Teachers should understand that "technology education must be holistic and technological activity undertaken by the students' needs to be authentic to technological practice and development with consideration given to social needs" (Turnbull 2013, p.49). Therefore, engaging students in authentic tasks will enhance teaching, and this is what happened when the teacher selected the robot as the unit that the students were to learn about because the robot has become one of the popular technologies that help people solve problems. Therefore, teaching this topic helped to engage students in authentic learning and thus helped them to understand what a robot is and how it can be programmed to do a certain task.

7. Enhancing students' technological literacy

   This is the overall goal of technology education. The teacher selected the topic and the content that helped to enhance students' technological literacy by engaging students in a task that represents technological knowledge as a main strand of the technology education concept. Students learned about the input, the system, and the output, which are among the purposes of teaching technology education that aim to enhance students' technological literacy.

What Else Would Be Good to Know, and How Can Teachers Find Out?

The Ministry of Education in Saudi Arabia has been working to improve education as a response to the new "Saudi Vision 2030" that is a part of the National Transformation Program. This program aims to push the country forward to be in an advanced and strong position among other advanced countries by developing a vibrant society, a thriving economy, and an ambitious nation. Also, the general aim of this strategy is to reduce the country's dominant dependence on oil. Science and technology will be a key enabler and driver of this transformation in Saudi Arabia, and preparing a new generation for this task has become imperative. This is unlikely to happen unless Saudi reviews its national curriculum, especially in terms of science and technology.

Some teachers are interested in reading about educational reforms that have been undertaken in many countries as they may provide some ideas and initiatives to improve education in those countries. This chapter highlights some ideas and information for those teachers who want to know about educational reform in Saudi and what strategy has been implemented to develop the country in general and education in particular. The new vision of Saudi 2030 opens the door for other teachers to contribute and make a change in the country. This chapter provides those teachers with a background of the status quo of technology education in our country. This will help them know what technology education means in Saudi and what kind of curriculum we teach in the classroom. As an educator and a researcher in the curriculum, I believe that integrating science and technology education in the classroom will be a key step toward improving the quality of teaching in Saudi Arabia in order to develop students' scientific and technological literacy.

Applying the model (Fig. 2.1) could be useful in improving the quality of teaching and learning through enhancing students' scientific and technological literacy. This will result in significant change in schools that foster millions of students who have been waiting for a modern curriculum and pedagogies that will help them to be effectively involved in the workplaces to create a prosperous nation.

This research can be accessed in the library of Canterbury University: my research is titled

"Technology Education in Saudi Arabia in Comparison with New Zealand: A study of Policy, Curriculum and Practice in Primary Education."

References

Al-Ghamdi, M. (2012). *Content evaluation of the developed primary science curriculum in Saudi Arabia*. (Master's thesis), Um-Alqura University: Makkah.

Al-Ghamdi, A., & Al-Salouli, M. (2013). Saudi elementary science teachers' beliefs: Teaching science in the new millennium. *International Journal of Science and Mathematics Education, 11*(2), 501–525.

Almutairi, A., Everatt, J., Snape, P., & Turnbull, W. (2014). Exploring the relationship between science and technology in the curriculum. *Australasian Journal of Technology Education, 1*, 49–63.

Bryman, A. (1988). *Quantity and quality in social research*. London: Unwin Hyman.

Cajas, F., & Gallagher, J. (2011). The interdependence of scientific and technological literacy. *Journal of Research in Science Education, 38*(7), 713–714.

Compton, V. (2004). *The relationship between science and technology: Discussion document prepared for the New Zealand Ministry of Education Curriculum Project*. Retrieved from www.tki.org.nzcurriculum/whats_happening/index_e.php

De Vries, M. (2001). The history of industrial research laboratories as a resource for teaching about science-technology relationships. *Research in Science Education, 31*(1), 15–28.

De Vries, M. (2012). International curricula requirements for making connections in science and technology education. In B. France & V. Compton (Eds.), *Bringing communities together: Connecting learners with scientists or technologists* (pp. 43–59). Rotterdam: Sense.

Fath-Allah, M. (2006). *Teaching technology in general schools* (1st ed.). Riyadh: Alsomaie House.

Gravemeijer, K., & Baartman, L. (2011). Science and technology education for the future. In M. De Vries, H. Keulen, S. Peters, & W. der Molen (Eds.), *Professional development for primary teachers in science and technology* (pp. 22–33). Rotterdam: Sense.

Guest, G., Bunce, A., & Johnson, L. (2006). How many interviews are enough?: An experiment with data saturation and variability. *Field Methods, 18*, 59–82.

Jones, A., & Carr, M. (1992). Teachers' perceptions of technology education: Implications for curriculum innovation. *Research in Science Education, 22*(1), 230–239.

Kipperman, D. (2006). Science and technology links in Israeli secondary schools: Do we have a reason to celebrate? In M. De Vries & I. Mottier (Eds.), *International handbook of technology education: Reviewing the past twenty years* (pp. 487–497). Rotterdam: Sense.

Lebeaume, J. (2011). Between technology education and science education: A necessary positioning. In M. De Vries (Ed.), *Positioning technology education in the curriculum* (pp. 75–86). Rotterdam: Sense.

Ministry of Education. (1995). *Technology in the New Zealand curriculum*. Wellington: Learning Media.

Ministry of Education. (2007a). *The New Zealand curriculum*. Wellington: Learning Media Limited.

Ministry of Education (2007b). Technology curriculum support: A package of documents and papers developed by the Ministry of Education to support schools and teachers with the implementation of the technology curriculum in the New Zealand Curriculum (2007). Retrieved from www.techlink.org.nz/curriculum-support

Ministry of Education. (2008). *Policy and strategy*. Retrieved July 7, 2012 from http://www.minedu.govt.nz/NZEducation/EducationPolicies/Schools/PolicyAndStrategy/PlanningReportingRelevantLegislationNEGSAndNAGS/TheNationalEducationGuidelinesNEGs.aspx

Ministry of Education (TKI). *Technology online: indicators of progression*. Retrieved from http://technology.tki.org.nz/Technology-in-the-NZC/Indicators-of-progression

Morgan, D. (1998). Practical strategies for combining qualitative and quantitative methods: Applications to health research. *Qualitative Health Research, 8*(3), 362–372.

Punch, K. F. (2005). *Introduction to social research: Quantitative and qualitative Approaches* (2nd ed.). London: SAGE.

Ryan, G., & Bernard, H. (2003). Techniques to identify themes. *Field Methods, 15*(1), 85–109.

Sparkes, J. (1993). Some differences between science and technology. In R. McCormick, C. Newey, & J. Sarkes (Eds.), *Technology for technology education* (pp. 25–36). Padstow: Addison-Wesley.

Turnbull, W. (2002). The place of authenticity in technology in the New Zealand Curriculum. *International Journal of Technology and Design Education, 12*(1), 23–40.

Turnbull, W. (2013). *The nature of conversation of primary students in technology education implications for teaching and learning*. PhD thesis, University of Waikato, Hamilton, New Zealand.

Chapter 3
The Localisation of Technology Education Curriculum in Botswana

Victor Ruele

Abstract This chapter explores the management of change from a British model of design and technology (D&T) curriculum to the Botswana model designed for senior secondary schools. The study employed the ADKAR (awareness, desire, knowledge, ability and reinforcement) change model, originally designed for business and industry as a diagnostic tool to assess the nature of change and use insights gained to identify gaps in the implementation and make proposals for more effective implementation. In terms of data collection, a multi-method approach was adopted which included questionnaires, individual and group interviews as well as document review. The bulk of the data were collected from in-service officers and teachers because of their role as change managers and implementers, respectively. A number of challenges are identified with the implementation of this curriculum as follows: a limited implementation strategy, a limited participation by key stakeholders, a weak coalition for change, a limited administrative support especially in terms of provision of resources and a limited teacher support system as well as weak reinforcement mechanisms to sustain the change. The ADKAR framework employed in this study was found to be a useful diagnostic tool to assess readiness across all levels of the Ministry of Education in terms of the capacity to implement a new curriculum.

The Questions Asked and Why They Are Important

This study investigated the experiences of the implementation of a new design and technology curriculum in Botswana, the capacity for change and the kinds of support needed to cope with the change. The goals were to identify and evaluate the necessary steps and strategies for managing the change and transition to ensure effective implementation in schools. So, the primary research question which

V. Ruele (✉)
University of Botswana, Gaborone, Botswana
e-mail: ruelev@mopipi.ub.bw

P J. Williams, D. Barlex (eds.), *Explorations in Technology Education Research*, Contemporary Issues in Technology Education,
https://doi.org/10.1007/978-981-13-3010-0_3

drove this inquiry was: *What is the nature of change from the Cambridge D&T curriculum to the Botswana model, and how are teachers being supported to implement the curriculum?*

As you will note, the main research question had two parts to it which I considered important to investigate from my perspective as a reflective practitioner and an educator. Firstly, having been a member of the task force which was involved in the development of this curriculum, I was concerned that ever since we moved away from the Cambridge curriculum to a national curriculum in 2000, there had never been any piloting of the new curriculum before full implementation. Furthermore, there had been no previous study to assess how this curriculum was being implemented in schools and the challenges presented. Only one study by Ndaba (1994) was found which attempted to address implementation issues at senior secondary level in the early 1990s when the Cambridge model was being introduced in Botswana. Also, in terms of implementation, there is little documentation available to teachers and in-service officers in the form of guidance materials. Secondly, I was interested in exploring how the teachers as implementers were being supported through the transition period. Studies on educational change management have underscored the importance of stakeholder participation especially the teachers in the change process, the need to understand the context for change, continuous administrative support and provision of resources and the need for continuous professional development of the teachers as key change agents in schools (Morrison 1998; Fullan 1993, 2001; Lachiever and Tardif 2002; Eggleston 1996; Atkinson 1990; Jones 2003). So in summary, the specific objectives of the study were to:

(a) Gain insights into the nature and scope of change from the Cambridge curriculum to the locally developed D&T curriculum for senior secondary schools in Botswana.
(b) Assess the general awareness and support for the new D&T curriculum among the various stakeholders involved with the implementation of the curriculum in Botswana.
(c) Identify and analyse key factors that facilitate or hinder the implementation of BGCSE D&T curriculum in senior secondary schools, and consider how these might be overcome to ensure effective implementation.
(d) Apply the findings of the study to make informed suggestions for appropriate change management capabilities and competencies needed for effective implementation of the BGCSE D&T curriculum.

Formulation of Research Questions

From this understanding, I then deemed it would be necessary to develop research questions which would enable me to assess the capacity to implement the curriculum from the perspective of the teachers as implementers. Harvard Business Press

(2003) underscores this notion that people closest to a situation can address the issues better than those in the management.

The objectives stated above led to the formulation of the following research questions:

(a) *What factors have necessitated change by the government of Botswana from the Cambridge D&T curriculum to a locally developed BGCSE D&T curriculum?*
(b) *What change management capabilities and competencies were developed by the sponsor to facilitate effective implementation of the planned change?*
(c) What *organisational factors have facilitated or inhibited effective implementation and management of change and transition in design and technology?*
(d) What *change management capability needs to be developed to facilitate effective implementation of the BGCSE D&T curriculum?*

The Design of the Study and Data Collection Methods

This next section explains and justifies the rationale behind the design and the methods used for data collection. The study was principally qualitative in nature because it was based on views and perceptions of those trying to implement the curriculum. However, for purposes of gaining greater insights, I employed both qualitative and quantitative methods. The quantitative approach involved a questionnaire survey which was administered to the teachers. The questionnaire survey consisted of both closed questions and open-ended questions. For the closed questions, a Likert scale was used to measure participants' views and perceptions on the issues. Open-ended questions were included to allow participants to elaborate their responses where necessary. The qualitative approach included semi-structured individual and group interviews with participants representing various levels of experience in order to provide a wide span of managers and stakeholders. Group interviews were used to provide a platform for the participants to air their views and allowed the researcher to probe issues to deepen understanding. Also used were insights gained from seminars, conferences and in-service training (INSET) workshops conducted for teachers. Participant selection was driven primarily by the research questions and the kinds of data required.

Having developed the research questions to be investigated, the next stage was to identify participants who could help answer those questions as well as the methods which would be used to collect data from those participants. Through data collection instruments designed around the above research questions, I would be able to capture the views and perceptions of the various stakeholders in terms of the organisational conditions under which the change took place and the capabilities that had been developed to facilitate successful implementation.

Exploration of Relevant Literature

This research was conducted against the backdrop of issues and concerns which have been highlighted in the Final BGCSE Evaluation Report (2009). The challenge of implementing the D&T curriculum at senior secondary level was compounded by limited research into this area. Much of the research done in Botswana has focused on the earlier years of schooling and addressed issues pertaining to the historical perspective of the subject, teachers' and students' perceptions and classroom experiences in the teaching and learning of D&T education (Ndaba 1994; Moalosi 1999; Fox 1988; Molwane 2003; Moalosi 2007; Gaotlhobogwe 2010). The lack of research in the subject area limits the efficiency of any curriculum review process. On the other hand, the researcher noted considerable research into this area in more developed countries, notably England, Australia, New Zealand and the United States (Atkinson 1990; Rasinen 2003; Williams 2007), but the contexts of these countries differ significantly to that of a developing country like Botswana (Ndaba 1994; Lewis 2000; Williams 2007). However, studies in these contexts provided insights which helped to lay the theoretical framework for this research project.

To explore and contextualise issues found in the literature, a pilot study was designed to collect baseline data which helped to focus the study. The researcher used concepts derived from the literature as theoretical framework for the research.

Figure 3.1 summarises the conceptual framework which guided the design of the pilot survey. As seen in Fig. 3.1, the research questions were developed around key issues or themes (shown **bolded**) which arose from the literature.

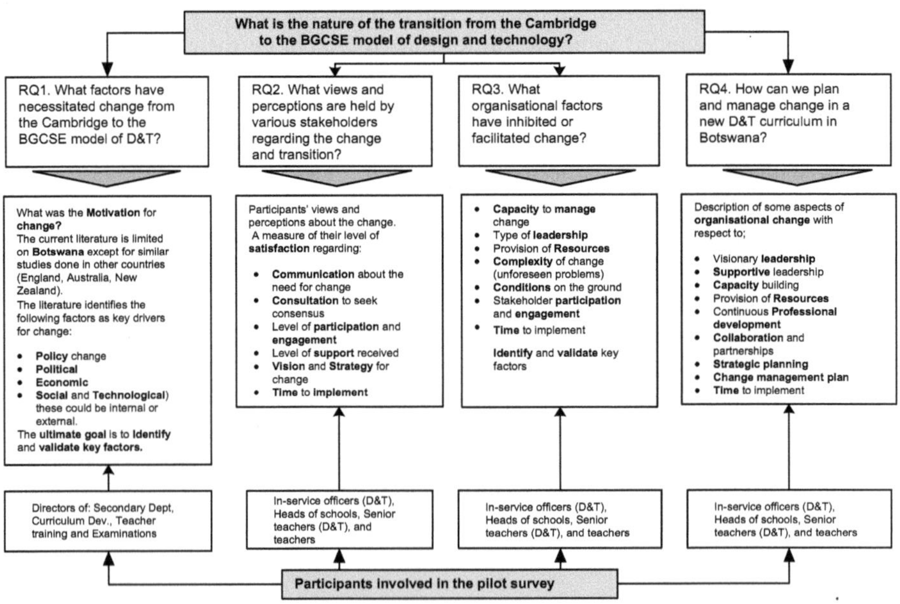

Fig. 3.1 Pilot survey conceptual framework

Exploration of Change Management Literature in Botswana

As already indicated in the initial literature review which I explored, I found that it was limited in addressing curriculum change within the context of Botswana. Therefore, I felt that it would be helpful if I could design a small, pilot study to explore some of these issues from the context of Botswana. I then used the findings from the pilot study to determine the direction and focus of my study. The pilot study was exploratory in nature because little was known about the area of research being investigated (Neuman 2011). The idea of designing a pilot study was to enable me to collect baseline data whose findings I could use to inform the scope and direction of a subsequent study, whereas initial literature findings were used to guide the formulation and focus of research questions for this initial research.

Exploration of Change Management Literature Review

From the literature review which I explored on change management, I identified many factors which affect change. Figure 3.2 summarises some of the summary of these key factors which arose from literature on change management.

The key findings in respect of this study are as follows: (a) The localisation of the curriculum was widely hailed as a move in the right direction; (b) it has been noted

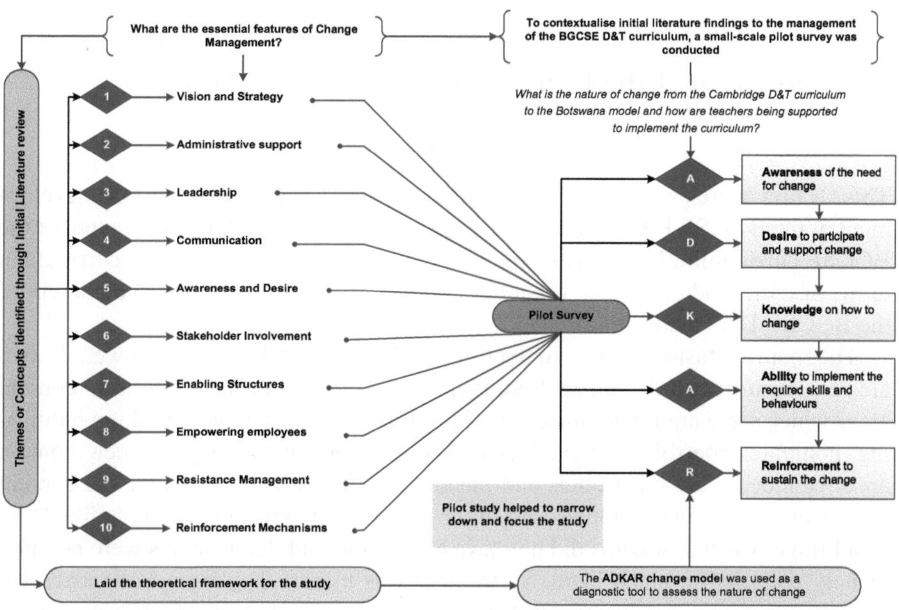

Fig. 3.2 A summary of key factors which arose from change management literature insights gained from the study

that some progress has been made in terms of the training of D&T teachers. Many of them possess the relevant qualifications and seem to be quite experienced. However, it seems that their training was limited especially in ICT, CAD/CAM and the technology content areas; (c) in terms of administrative support and commitment, the general observation is that the government as the main sponsor did not commit fully to the project. For instance, there remains the unfinished business on the part of the government in terms of provision of equipment for ICT, CAD/CAM and control technologies particularly pneumatics. It was noted that the government provided mainly the physical infrastructure but did not provide the necessary equipment to support the teaching and learning of the content areas mentioned above.

Changes to the New D&T Curriculum

The aim of this question was to assess whether the new curriculum had included any new content areas and how the teachers were being supported to cope with new content. It was addressed to the curriculum developers, teachers and in-service officers. Unlike its predecessor which was predominantly craft based, the BGCSE D&T curriculum includes new content areas, namely, electronics, pneumatics and computer-aided design and manufacture. The inclusion of these content areas was to align the curriculum with the country's vision of moving away from the traditional agro-based economy to more of an industrial focus (Republic of Botswana 1994).

A Comparison of the Cambridge D&T Curriculum with the BGCSE

This section shares with the reader the changes that were brought into the development of the BGCSE D&T curriculum. First I present the structure and content of the BGCSE curriculum. Next I will explain the difference between the two curricula in terms of focus and emphasis. Figure 3.3 presents the basic structure and content of the BGCSE D&T curriculum.

The main inclusions to the content of the BGCSE D&T curriculum were in two areas: **communication** and **technology**. Under communication, the new content areas which were not taught under the Cambridge curriculum were desktop publishing, computer control and CAD/CAM. Whilst under technology, the new content areas included were electronics and pneumatics. These new technological content areas were not strongly emphasised in the Cambridge curriculum. The rationale by Cambridge was that schools did not have equipment and that teachers were not adequately prepared to teach them. So the fear was that, if they were to be assessed, students would be disadvantaged. Whereas in the case of the BGCSE programme, these content areas were deemed important to enhance knowledge and skills as well

Syllabus Component	Content Areas
1. Health and Safety	Safety Precautions and First Aid.
2. Materials	Timber; Manufactured boards; Metals; Plastics; Optional materials; Adhesives; Abrasives; Fixings; Fittings and Finishes.
3. Communication	Graphics and Information Technology (Desktop publishing; Computer control; Draw programs; CAD/ CAM).
4. Design	Design Process; Marketing; Promotion; Costing and Pricing production.
5. Technology	Energy; Structures; Mechanisms; Electronics and Pneumatics.
6. Tools and Processes	Measuring and Marking out; Saws and Sawing; Planes and Planing; Files and Filing; Drills and Drilling; Chisels and chiselling; Shears and shearing, Forming, Moulding and Casting; Turning and Milling; Joining and Fabricating; Holding and Assembling tools; and Finishing materials.

Fig. 3.3 Basic structure of the BGCSE D&T curriculum. (Source: The BGCSE D&T Teaching Syllabus (2000))

as to keep up with the pace of technological development (Ministry of Education 2000). Content areas such as health and safety, materials, design and tools and processes have always been part of the Cambridge curriculum. The only thing the curriculum developers did in Botswana was to increase breadth and depth of content coverage. This results in not only a congested curriculum but one that is still deeply rooted in the traditional crafts. In fact some have expressed the view that there has not been much change in approach to the way the subject used to be taught under Cambridge ("The BGCSE programme is fairly new people want to stick to the old system. We're still in a transition"; Source: Set 4). Still on the same issue, one officer suggests that: "teachers continued to work along traditional lines" (Source: Set 3).

How This Might Be Used to Improve Teaching and Learning?

As already indicated at the introductory section, my study set out to investigate a range of issues relating to the processes and challenges of implementing a newly developed D&T curriculum for senior secondary schools in Botswana. The ultimate goal of the study was to be able to apply the findings towards improving classroom practice. In particular, the researcher was interested in examining how the new curriculum was being implemented through the experience of teachers as change agents. The study identified a range of factors which contributed to the partial implementation of the BGCSE D&T curriculum. Overall, the major issues which were raised by the teachers concerning the implementation of the curriculum included:

- A "crowded curriculum" which was found to be a major challenge for many teachers.
- Teachers expressed the need for upskilling or professional development in technology education and, in particular, electronics, pneumatics and computer-aided design and manufacture (CAD/CAM).
- The difficulty of resourcing the equipment needed to implement the above-mentioned technology content areas.
- Implementing the curriculum in an integrated (holistic) manner was one of the major challenges for many teachers.

Also one of the key findings of the study was that of teachers' limited capability to deliver some new content areas of the curriculum such as electronics, pneumatics and CAD/CAM. Limited teacher capability was deemed a worrying factor considering that many D&T teachers were trained overseas in preparation for the localisation exercise. As a result of this deficiency, it was reported that students' performance in the subject was poor, and their enrolment in the subject was also declining. In the light of these findings, teachers can be supported to deliver the curriculum in the following ways.

There is a need to develop some guidance materials especially for the most affected content areas of the curriculum. For example, in the early 1990s in the United Kingdom, a project known as Nuffield Foundation was initiated to develop a wide range of support materials for teachers and students.

> Nuffield Design & Technology gathered these various insights into something coherent that all teachers could share and use in their teaching across the new subject. The project gave teachers a voice, listened to that voice, and used this to produce useful approaches to teaching and learning. (http://www.nuffieldfoundation.org/nuffield-design-technology)

In Botswana, the development of such materials should be informed by the ongoing revision of the BGCSE curriculum. An integrated framework should be developed in conjunction with the office of Curriculum Development and Evaluation (CD&E) to provide more specific guidance on the implementation of such a system.

- It has been noted that BGCSE D&T syllabus is congested and cumbersome to teach effectively. Consequently, teachers resorted to the teaching of theory at the expense of practical work. The need, therefore, is a syllabus breakdown to enable it to be offered in an integrated manner.
- The development of subject leadership. Leadership refers to a guiding authority that determines and enables change to occur (Hiatt and Creasey 2012). The notion of leadership in change management is to set direction, solve problems, create a vision and lead people in an organisation to a better future. It was found that the current in-service officers have not been trained sufficiently to deliver professional development programmes for teachers in D&T. One of the key

issues about change management is the development of leadership and management skills (Kotter 1996; Hiatt and Creasey 2012). As part of a leading coalition, in-service officers and senior teachers act as change managers and mentors; as such they need to be empowered with the knowledge and skills on how to coach others through change. To manage the implementation of the curriculum at school level, there is a need to identify subject leaders for different content areas of the curriculum. The criteria should be based on teachers' strength and interest in the specialist area. Once identified, these teachers should be given appropriate training in their specialist areas of the curriculum such as electronics, pneumatics and computer-aided design and manufacture. The ultimate goal is to achieve some parity in capability across all the schools.

Furthermore, it was reported in the study that in-service support is limited both at school and regional levels. With this in mind, this calls for:

- A skills audit for all the teachers to assess their skill level: The skills audit will inform training needs for teachers. This will need a coordinated approach by all the stakeholders.
- An extensive programme of professional development to be developed for teachers to strengthen in-service training and ongoing professional development: The department of Teacher Training and Development (TT&D) has been tasked with this mandate. However, TT&D does not have the capacity to deliver in-service training. They do not have enough personnel on the ground. This is evidenced by lack of officers in some regions. This problem can be alleviated by engaging lectures of Molepolole College of Education (MCE) and the University of Botswana. A mutual arrangement could be made whereby TT&D offers financial support, whilst the two institutions offer the human resource. Through a collective and sustained effort, these interventions will enable teachers to gain confidence in all aspects of the new curriculum and lead to improved student learning.
- Also since schools are organised into clusters, this arrangement has the potential to greatly increase the likelihood that changes in practice will be sustained. In fact, as was observed by Welch and Mueller (2003), this collaborative approach provides an opportune moment for teachers to engage one another in tackling issues pertaining to new subject matter and to engage in innovative work aimed at curriculum reform. Indeed, such a forum provides opportunities for teachers to form "communities of practice" that encourage them to reflect on the content and contexts of their pedagogy. Likewise Fullan (2001) argues that: "good ideas come from talented people working together" (p. 175).
- Finally, in view of the fact that it would take time to develop the necessary capacity to implement a school-based support system, it would not be advisable to discard the regional in-service support system because some regions like the southern have well-equipped education centres. Such a centre can be used by schools which are not well resourced. Moreover, the centres could serve as a meeting point for teachers across the schools to deliberate on issues of common interest such as curriculum planning, development of resource materials and networking.

Conclusions

Through this chapter, an understanding of the complexity of managing curriculum change, the need to understand the context for change, the continuous administrative support and provision of resources and the need for continuous professional development of the teachers as key change agents in schools have been established. From this understanding, it is evident that teachers play an important role in the successful implementation of a new curriculum, and therefore change management capability is a pertinent issue for education (Morrison 1998; Fullan 2001; Whitaker 1993). The study has established that if change agents are to be successful especially teachers and in-service officers, they should be well supported with training and coaching. Usually the demands of implementing change, together with a need to keep the day-to-day requirement of the job going, mean that everything gets done in a rush, without pausing to review, develop or integrate. Hence, mentoring, reviewing and feedback mechanisms need to be established to help the change process and to also build ongoing capability (Cameron and Green 2012; Kotter 1996).

The difficulties faced in Botswana in moving from an existing syllabus to a new one and the challenges thereof such as lack of up-to-date equipment, maintenance of existing equipment and teacher knowledge and skill in using such equipment are not unique to Botswana. However, as was experienced in England and New Zealand, implementing curriculum change is an enormous task; the government alone cannot effect change; it is a collaborative effort. To support collaborative effort, Chikasanda (2010, p. 16) notes that: "the political will in the whole process played a significant role as shown by the establishment of a Ministerial task group" (also see Jones 2003; Atkinson 1990; Eggleston 1996). Therefore, what is needed going forward is a shared vision, creation of a strong and credible coalition, creation of enabling environment, provision of resources, continuous monitoring and professional development of all those involved in the change process.

References

Atkinson, S. (1990). Design and technology in the United Kingdom: Historical perspective. *Journal of Design & Technology Education, 2*(1), 1–12.

Cameron, E., & Green, M. (2012). *Making sense of change management: A complete guide to the model, tools and techniques of organisational change* (3rd ed.). London: Kogan Page Limited.

Chikasanda, V. K. M. (2010). *An investigation of the development of students 'and teachers' perceptions towards technology: A framework for reconstructing technology education in Malawi.* (Unpublished PhD). University of Waikato, New Zealand, Centre for Science and Technology Education Research: University of New Zealand.

Eggleston, J. (1996). *Teaching design and technology* (2nd ed.). Buckingham: Open University Press.

Fox, R. (1988). *A report to the ministry of education, Botswana on completion of a consultation and in-service teaching tour*. Gaborone: Government Printers.

Fullan, M. (1993). *Change forces: Probing the depths of educational reform*. London: The Falmer Press.
Fullan, M. (2001). *The new meaning of educational change* (3rd ed.). London: Routledge Falmer.
Gaotlhobogwe, M. (2010). *Attitudes to and perceptions of design and technology students towards the subject: A case of five junior secondary schools in Botswana.* (Unpublished PhD). University of Wales Institute.
Harvard Business Press. (2003). *Managing change and transition*. Boston: Harvard Business School Publishing Corporation.
Hiatt, J. M., & Creasey, T. (2012). Change Management. In *The people side of change*. Colorado: Prosci Learning Centre Publications.
Jones, A. (2003). The development of a national curriculum in technology for New Zealand. *The Journal of Technology and Design Education, 13*, 83–99.
Kotter, J. P. (1996). *Leading change*. Boston: Harvard Business School Press.
Lachiever, G., & Tardif, J. (2002). Fostering and managing change and innovation. In Frontiers in education, 2002, FIE 2002, 32nd Annual (Vol. 2).
Lewis, T. (2000). Technology education and developing countries. *International Journal of Technology and Design Education, 10*(2), 163–179.
Ministry of Education. (2000). *Botswana general certificate of secondary education teaching syllabus: Design and technology*. Gaborone: Curriculum Development and Evaluation.
Moalosi, R. (1999). Historical perspective of traditional technical subjects in Botswana. *The DATA Journal of Design and Technology Education, 4*(2), 159–160.
Moalosi, D. R. (2007). *The impact of socio-cultural factors upon human-centred design in Botswana* (Unpublished PhD). Queensland University of Technology, Australia.
Molwane, O. B. (2003). *Assessing students' capability in design and technology education in junior secondary schools in Botswana.* (Unpublished PhD). Goldsmiths University of London, Department of Design; Goldsmiths University of London.
Morrison, K. (1998). *Management theories for educational change*. London: Sage Publications Company.
Ndaba, N. (1994). *The effects of the shift from the traditional craft subjects to design and* technology – *the Botswana experience.* IDATER Conference, Loughborough University.
Neuman, W. L. (2011). *Social research methods: Qualitative and quantitative approaches* (6th ed.). Boston: Pearson Education.
Rasinen, A. (2003). An analysis of the technology education curriculum in six countries. *The Jounral of Technology Education, 15*(1), 31–47.
Republic of Botswana. (1994). *The revised national policy on education*. Government paper No. 2. Gaborone: The Government Printer.
Republic of Botswana. (2009). *Final BGCSE evaluation report*. Gaborone: Department of Curriculum Development and Evaluation.
Welch, M., & Mueller, A. (2003). In the shoes of a student: Professional development in a classroom context. *The Journal of Design and Technology Education, 8*(2), 91–100.
Whitaker, P. (1993). *Managing change in schools: Developing teachers and teaching*. Buckingham: Open University Press.
Williams, P. J. (2007). A global curriculum: Design and technology in the international baccalaureate. *Design and Technology Education: An International Journal, 12*(3), 47–57.

Chapter 4
What Children Are Supposed to Learn in Primary Technology Education

Eva Björkholm

Abstract The aim of this chapter is to shed light on a specific and fundamental issue for teachers in primary technology education – What do students know when they achieve the knowledge goals for technology education? In order to design teaching, and assess and support students' development of knowledge, teachers need this kind of detailed knowledge. In the studies described, the capability to evaluate the fitness for purpose of technical solutions and to construct a specific linkage mechanism was explored in two technology classrooms with students aged 6–8 years. The results suggest aspects on which the teaching has to focus, in order to make the intended learning possible.

The Questions I Asked and Why They Are Important

One of the purposes of the Swedish subject of technology is to develop students' technical knowledge in a range of areas. However, it is not clear from the syllabus what the technical knowledge is made up of and how this knowledge can be developed. These things are not often problematized when teachers are planning for learning; instead, the questions of how teaching should be organized come to the fore. What the subject-specific knowledge is made up of is often taken for granted, and it has therefore not been discussed and examined (Ellis 2007). But this issue is far from obvious. Even teachers with deep knowledge of the subject may have difficulties in articulating what students really know when they have achieved the knowledge goals. Often, this kind of teachers' professional knowledge is lacking, even though it also exists as tacit knowledge. However, in order to plan teaching that enables students' learning, and to support students' learning as well as evaluating teaching and learning, teachers need detailed knowledge about what the students know when they achieve the set goals.

E. Björkholm (✉)
Royal Institute of Technology, Stockholm, Sweden
e-mail: evabjork@kth.se

P J. Williams, D. Barlex (eds.), *Explorations in Technology Education Research*, Contemporary Issues in Technology Education,
https://doi.org/10.1007/978-981-13-3010-0_4

For primary technology education, the teachers' task of designing activities that enable students to develop technology-specific knowledge is even more complicated as primary technology, in contrast to many other school subjects, lacks a common base of experience of teaching and assessment (cf. Stein et al. 2007). In the past, technology teachers have largely focused on technology as an opportunity for students to experience practical work. This means that technology teaching has not, or only to a small extent, explicitly focused on what the students are expected to learn (Siraj-Blatchford and MacLeod-Brudenell 1999; Swedish Schools Inspectorate 2014). Thus, there is a need for a defined body of knowledge for planning, teaching and evaluating student knowledge in primary technology education (Jones and Moreland 2003; Jones et al. 2013).

The purpose of my doctoral research was to contribute to the professional knowledge base of technology teaching in the early school years by developing and describing knowledge in technology. The subject-specific knowledge which was explored was the ability to evaluate technical solutions' fitness for purpose and knowledge related to the construction of a linkage mechanism. These abilities were chosen since they are both highlighted in the Swedish technology syllabus, and the choice was also based on teachers' experiences of something being difficult to teach. In addition, constructing mechanisms is a frequent activity in primary school, involving certain difficulties for young children.

The questions which guided the research were:

- What does it mean to be able to evaluate technical solutions and to construct a specific linkage mechanism in primary school?
- What do the children need to learn in order to develop these abilities?

How I Tried to Answer the Questions

Data were mainly generated through two so-called learning studies conducted in primary schools. Learning study is a classroom-based research approach with a high degree of intervention and collaboration between a group of teachers and the researcher (Marton and Pang 2006). The research is focusing on specific objects of learning, that is, on something that the students are expected to learn.

In the first learning study, focusing on the ability to evaluate the fitness for purpose of technical solutions, a group of four teachers collaborated with the researcher. In learning study two, two teachers participated focusing on constructing a linkage mechanism allowing for transferring and transformation of movement. The choices of the specific objects of learning were based on the teachers' previous experiences of something being difficult to teach, as well as content highlighted in the syllabus. After selecting the objects of learning, the next step was to design a pretest to map students' prior knowledge. Considering the result of the pretest, the lesson was designed on the basis of a specific theory of learning, the variation theory (Marton and Booth 1997; Marton et al. 2004), and then carried out by one of the teachers with a group of students. The lesson and the students' outcomes were then analysed, and based on this information, the lesson was revised. In the next cycle, the lesson was carried out with another group of students, and the procedure was repeated (Pang and Marton 2003). Both learning studies were conducted as a process of four cyclically repeated steps, investigating the most powerful way to teach the specific object of learning (Marton and Pang 2006). The two learning studies were conducted with a total of 49 pupils in grades 1 and 2 (7–8 years old) and 49 pupils in preschool class and grade 1 (6–7 years old), respectively.

Video recording was chosen for generating data, since the pupils' technical knowledge was understood to comprise skills and understanding as integrated forms of knowledge, that is, without separating thinking and doing. Moreover, the pupils' actions in relation to artefacts were focused upon. The empirical material generated in learning study one comprised video-recorded pretests in the form of conversations with pupils in pairs, focusing on familiar technical solutions for opening and closing, as well as video-recorded lessons. The base data of learning study two were video-recorded pre- and post-test concerning pupils' constructing a model of the linkage mechanism as well as the final models.

The transcribed video recordings were analysed using phenomenographic analysis (Marton 1981, 1994). The analysis resulted in categories describing pupils' qualitatively different ways of knowing the object of learning. Pupils' physical and verbal actions were analysed and understood as ways of knowing as a whole. The categories were ordered in a hierarchy and then analysed in terms of the aspects of the phenomenon that were focused on (Marton and Booth 1997). The differences between these aspects were seen as critical aspects, that is, aspects necessary for the pupils to discern in order to develop a more complex way of knowing.

What I Found Out

The findings from the first study describe different ways of knowing in terms of discerning functions related to different types of users, as well as aspects of the construction in order to realize functions.

The second study identified technical knowing as a specified analysis of the construction in terms of location and separation of joints in relation to different functions. These findings were then used to identify technical knowing in video data generated in another primary technology classroom working on construction. The results suggest that knowledge of specific objects of learning related to the evaluation and construction of technical solutions is partly generalizable. In addition, the specified knowledge concerning the meaning of the object of learning generated during the learning study process was described. This knowledge is suggested to be an important product of learning studies. The result of the two studies, as well as a follow-up study, will be described in more detail in the following section.

Knowing to Evaluate the Fitness for Purpose of Technical Solutions

Through the phenomenographic analysis, four qualitatively different ways of knowing to evaluate technical solutions' fitness for purpose were identified in the group of students. The categories are logically related to each other in that the first category is included in the next; thus the first category describes the least complex way of knowing. The categories describe different ways of knowing technical solutions' fitness for purpose as:

- *Effectiveness for me* – fulfilling functions based on own needs and everyday situations

In this category the students related the fitness for purpose only to their own use of the artefacts.

- *Effectiveness for others* – fulfilling functions based on others' needs and situations

In addition to their own use, the students evaluated the artefacts in relation to other people's uses.

- *Construction dependency* – how efficiently aspects of the construction help to realize the function

In this category the students also analysed the construction of the technical solution in terms of material, form and components.

- *Technical efficiency* – how efficiently commonly agreed technical solutions realize the function

This category contained the most complex way of understanding, meaning that common technical solutions, for example, the hinge, were identified and compared to other solutions in terms of how efficiently they realize specific functions.

Based on the categories described above, critical aspects were identified. These are aspects necessary for the pupils to discern in order to be able to evaluate the fitness for purpose of technical solutions:

(a) Functions related to needs of other users than me
(b) Material, form and components and how they contribute to fulfilling the functions. Linking main and secondary functions to the corresponding key components of the construction
(c) Common technical solutions and how they are constructed in terms of interaction between key components

How to Construct a Linkage Mechanism

In this study, the meaning of being able to construct a specific linkage mechanism (more specifically to assemble parts into a technical solution allowing for transferring and transforming motion) was examined (see Fig. 4.1).

Through the phenomenographic analysis, four qualitatively different categories emerged describing students' ways of experiencing/knowing to construct the mechanism. As in the first study, the categories describing the less complex ways of knowing were integrated in the more complex ways. The four categories can be summarized as constructing the specific mechanism as:

- *Testing components* – the driving movement is directed at the same point as the resulting movement, and the components are tested to realize the function (Fig. 4.2).

This way of knowing focused on testing the components and putting both linkages in the same point, resulting in only one fixed joint which the linkages could rotate around.

- *Making space for machinery* – the driving movement is directed at a different point than the resulting movement, thus creating space for both input and output as well as the mechanism for transferring movement (Fig. 4.3).

In this category, the linkages were attached in two fixed joints, creating a model with linkages that could not move.

- *Controlling machinery* – the components are attached to a fixed and a moving joint. The distance between input and output is short and contains few intermediaries, and the driving movement has the same direction as the resultant (Fig. 4.4).

This way of constructing the linkage mechanism resulted in a model which is similar to the transfer of movement when handling a marionette.

Fig. 4.1 The linkage mechanism

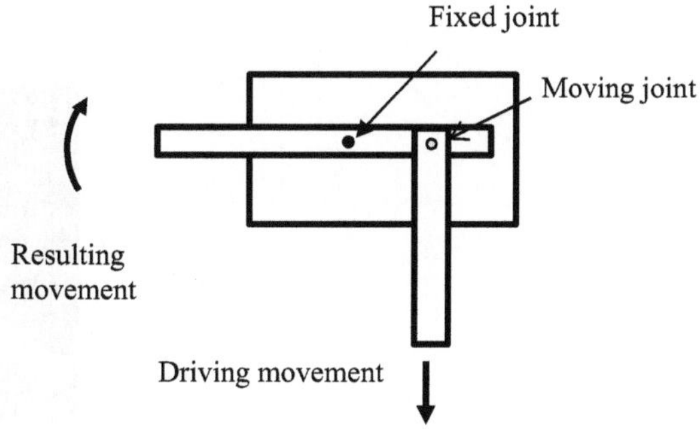

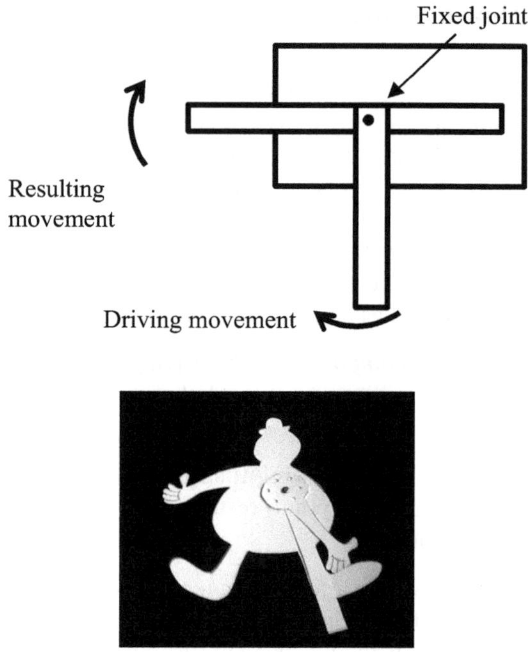

Fig. 4.2 The way of constructing categorized as testing components

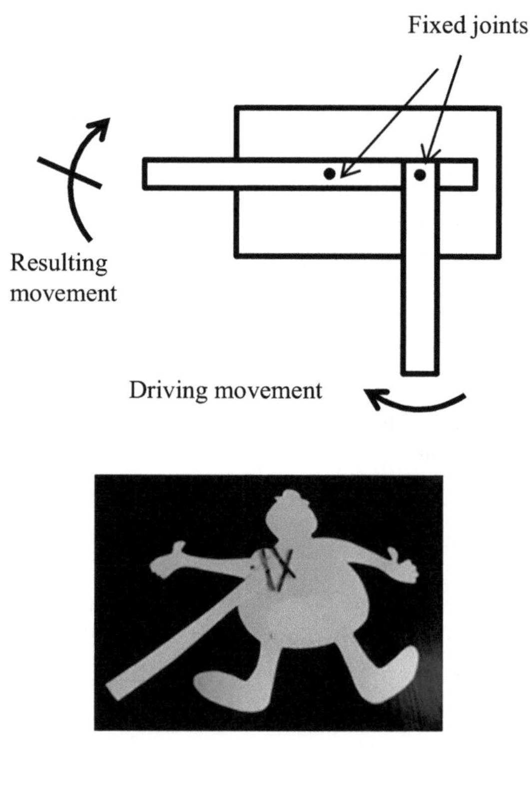

Fig. 4.3 The way of constructing categorized as making space for a machinery

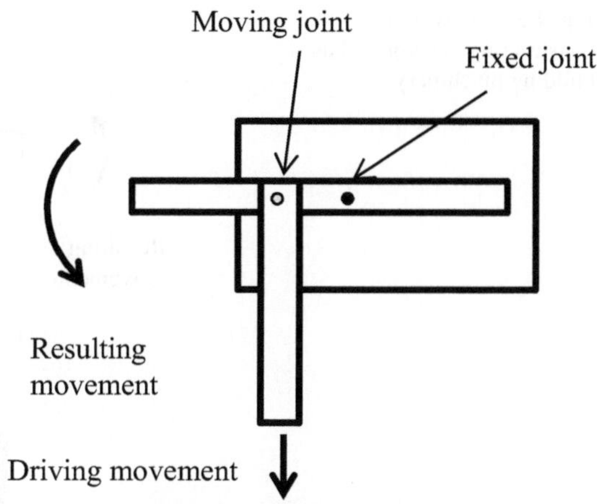

Fig. 4.4 The way of constructing categorized as controlling machinery

- *Building machinery* – a distinction between the fixed and moving joints is made, and the moving joint is placed on the opposite side of the resulting movement (Fig. 4.5).

In this category, the model contained the two different types of joints allowing for transferring and transformation of movement.

It was necessary for the students to be able to discern the following critical aspects in order for them to be able to construct the linkage mechanism:

(a) Number of joints
(b) Separating the moving joint from the fixed joint
(c) The position of the moving joint in relation to the resulting movement

During the learning study process, the object of learning was progressively unpacked, in terms of more and more specified and nuanced knowledge concerning the critical aspects. This specification gave rise to a need to articulate the critical aspects in terms of the way of constructing the mechanism. Therefore, the concepts of attack point and pivot point were replaced by the concepts of moving joint and fixed joint, respectively.

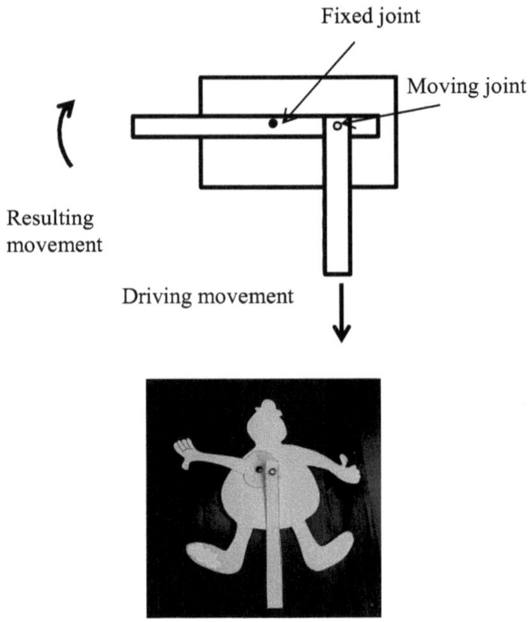

Fig. 4.5 The way of constructing categorized as building machinery

Identified Technical Knowing in the Primary Technology Classroom

The purpose of this study was to use these findings concerning technical solutions in order to identify technical knowing in a new teaching context. In this way, generalizable aspects of technical knowing in the primary technology classroom were explored. In the observed classroom, the students (7–8 years) were planning, sketching and building a model of a town. The findings suggest that the previous categories of technical knowing could be used in different ways. In relation to the evaluation of technical solutions, technical knowing was identified and specified in terms of:

- Realizing the function of representation as a sketch and as a model
- Relating the sketch to the model
- Choosing materials based on functional properties
- Building stable structures
- Adding components related to secondary functions

Concerning the construction of technical solutions, students' actions in relation to processing and joining materials were frequently observed in the video material. The identified categories of technical knowing were:

- Selecting and using tools for cutting different materials
- Making permanent joints in different materials
- Making permanent joints related to the size of the contact surface
- Making flexible joints

The ways of constructing were mostly focused on making permanent joints, except on one observed occasion when a flexible joint was made. All the identified categories of technical knowing were understood as less complex than those identified in the study focusing on the construction of a linkage mechanism. Therefore, the former categories could not be used to specify technical knowing in the technology lessons. Instead, these categories were added to the previously identified categories concerning the construction of technical solutions.

How This Might Be Used to Improve Teaching and Learning

The results of the studies consist of descriptions of technical knowing and technology content in relation to two specific learning objects in primary technology education. In addition, aspects have been identified that students need to discern in order to be able to develop these knowings. The understanding and articulation of what technical knowing and technology content consist of can be useful for teachers in their professional practice in several different ways, and I will now discuss this more in detail.

In *overall curriculum planning*, this kind of knowledge is important since it contributes to in-depth understanding of students' different ways of knowing and the difficulties they may encounter in learning it. Some guidelines are provided for how curriculum planning that includes a clear progression in learning can be designed. In addition, on the basis of this understanding, the objectives for students' learning can be explicitly formulated. Based on the results of the study focusing on simple mechanisms, it seems that constructions using just one fixed joint could be relevant to start with, followed by linked levers with the input and output movement having the same direction.

In *assessment for learning*, this knowledge is crucial for teachers, since it focuses on students' different ways of knowing in terms of complexity. This offers opportunities for teachers to identify the ways of knowing in a group of students and, based on this, organize teaching that gives students maximum opportunity to learn what is intended. The articulation of this professional knowledge also helps teachers to clarify and concretize in communication with students what they need to learn in order to achieve the learning goals. Moreover, in the communication between teachers regarding assessment, this articulated knowledge is crucial in order to be able to concretely discuss in detail what the knowing that is assessed consists of and in that way open up for creating possibilities for more equitable assessments.

Regarding *teaching methods in lessons*, the obtained knowledge is crucial for designing teaching that gives students opportunities to develop the intended knowledge. This is about how the content itself is presented and treated within the teaching. In the described studies, the variation theory serves as a theoretical tool for designing the teaching. In the design, implementation and evaluation of teaching, the variation theory or other theories of learning can be used. However, since my results were based on variation theory, I will give a brief description of this theoretical framework, followed by an example of how the aspects identified in the study can be used in the design of teaching in order to give students possibilities to learn.

According to *variation theory*, learning something is seen as a change in the learners' ability to discern aspects of this something (Marton 2015). From this follows that, in order to make learning possible, these aspects must be possible to be discerned by the learner. In order to discern them, students must experience these aspects as so-called dimensions of variation. For example, one would not be able to discern the aspect of colour if there was only one colour in the world. In the study concerning the construction of a linkage mechanism, an aspect that was critical for the group of students was to separate the moving joint from the fixed joint. In order to make the students discern this aspect, they were guided to try and construct the mechanism with two fixed joints (resulting in no movement) and compare it to constructing with one fixed joint and one moving joint (resulting in movement). By comparing the results in terms of movement or not, and clearly showing and discussing how the two different ways of constructing contributed to the difference, the students were given possibilities to learn how to construct the mechanism. During the lessons, the teachers and researcher discovered that the teaching material that students were given for constructing the mechanism model, which was paper and paper fasteners, created difficulties for students to discern the moving joint because of the thin material. This shows that the materials given to students when constructing contributes to what it is possible for them to learn.

A relevant objection to the usefulness of the obtained results may be how these ways of knowing and critical aspects which were identified in a few student groups can be used in other classrooms, in other schools and with other teachers. Although the identified critical aspects are specific to some student groups, teachers and teaching contexts, research shows that they can still be communicated to other teachers and used as resources in the work of other student groups (Kullberg 2010; Runesson and Gustafsson 2012). These aspects may have a guiding function in the planning and implementation of teaching in new student groups. However, in order to do this in a way that allows students to discern the critical aspects and thereby develop the knowledge, teachers need ideas about how the content will be presented and treated in the teaching (Nuthall 2004). Such a possible theoretical tool for planning and analysing teaching can be offered by the variation theory.

Several factors referring to the nature of the studies indicate that these results may be useful to teachers in technology education. Firstly, the teaching content that was investigated was based on teachers' own problems and questions in their teaching practice (Bulterman-Bos 2008; Korthagen 2007; Runesson 2011). Secondly, knowledge has been developed in collaboration with teachers in a teaching context where teachers' tacit knowledge has been involved (Bulterman-Bos 2008; Carlgren 2012).

What Else Could Be Good to Know, and How Could Teachers Find Out?

Teaching is always about *something* – some specific knowledge that the students are supposed to learn. By exploring this knowledge, and how it is constituted in the technology teaching practice, teachers gain specific insights which could be described as a combination of knowledge about content, what students need to experience and how to enact this in teaching. The approach described in the thesis could be used by teachers to further explore knowledge related to teaching and learning of constructing the same type of mechanism but with other groups of students or with students of other ages than those described here. Moreover, what students need to discern in order to construct other types of mechanisms taught in the primary curriculum, such as pulley systems, cams and followers and gear trains, could be explored in the same way.

The teacher of course needs to become familiar with the technical knowledge in order to analyse students' ways of constructing. Based on these insights, the teaching is planned and designed using variation theory. Students' learning is analysed, and the lesson is revised in order to enhance the specific learning. The results of these kinds of studies would then be reported in a professional journal in order to share with other teachers. Knowledge produced in a learning study could, however, not be seen as a canon for best practice or a detailed instruction for effective teaching, but a description of the aspects that have been found critical for a specific group of learners' learning of a specific object of learning.

References

Bulterman-Bos, J. A. (2008). Will a clinical approach make education research more relevant for practice? *Educational Researcher, 37*(7), 412–420.

Carlgren, I. (2012). The learning study as an approach for "clinical" subject matter didactic research. *International Journal for Lesson and Learning Studies, 1*(2), 126–139.

Ellis, V. (2007). Talking subject knowledge seriously: From professional knowledge recipes to complex conceptualizations of teacher development. *The Curriculum Journal, 18*(4), 447–462.

Jones, A., & Moreland, J. (2003). Developing classroom-focused research in technology education. *Canadian Journal of Science, Mathematics and Technology Education, 3*(1), 51–66.

Jones, A., Buntting, C., & de Vries, M. J. (2013). The developing field of technology education: A review to look forward. *International Journal of Technology and Design Education, 23*(2), 191–212.

Korthagen, F. A. J. (2007). The gap between research and practice revisited. *Educational Research and Evaluation: An International Journal on Theory and Practice, 13*(3), 303–310.

Kullberg, A. (2010). *What is taught and what is learned: Professional insights gained and shared by teachers of mathematics*. Doktorsavhandling. Göteborg: Göteborgs universitet.

Marton, F. (1981). Phenomenography – Describing conceptions of the world around us. *Instructional Science, 10*(2), 177–200.

Marton, F. (1994). Phenomenography. In T. Husén & T. N. Postlethwaite (Eds.), *The international encyclopedia of education* (2nd ed., pp. 4424–4429). Oxford: Pergamon Press.

Marton, F. (2015). *Necessary conditions of learning*. London: Routledge.

Marton, F., & Booth, S. (1997). *Learning and awareness*. Mahwah: Lawrence Erlbaum.

Marton, F., & Pang, M. F. (2006). On some necessary conditions of learning. *Journal of the Learning Sciences, 15*(2), 193–220.

Marton, F., Runesson, U., & Tsui, A. B. M. (2004). The space of learning. In F. Marton & A. B. M. Tsui (Eds.), *Classroom discourse and the space of learning* (pp. 3–40). Mahwah: Lawrence Erlbaum.

Nuthall, G. (2004). Relating classroom teaching to student learning: A critical analysis of why research has failed to bridge the theory-practice gap. *Harvard Educational Review, 74*(3), 273–306.

Pang, M. F., & Marton, F. (2003). Beyond "lesson study": Comparing two ways of facilitating the grasp of some economic concepts. *Instructional Science, 31*(3), 175–194.

Runesson, U. (2011). Lärares kunskapsarbete: exemplet learning study. In E. Forsberg (Ed.), Lärare som praktiker och forskare: om praxisnära forskningsmodeller. *Forskning om undervisning och lärande 5,* 7–17.

Runesson, U., & Gustafsson, G. (2012). Sharing and developing knowledge products from learning study. *International Journal for Lesson and Learning Studies, 1*(3), 245–260.

Siraj-Blatchford, J., & MacLeod-Brudenell, I. (1999). *Supporting science, design and technology in the early years*. Buckingham: Open university press.

Stein, S. J., Ginns, I. S., & McDonald, C. V. (2007). Teachers learning about technology and technology education: Insights from a professional development experience. *International Journal of Technology and Design Education, 17*(2), 179–195.

Swedish Schools Inspectorate. (2014). *Teknik – gör det osynliga synligt*. Om kvaliteten i grundskolans teknikundervisning. Rapport 2014:4. [Technology – Make the invisible visible. About the quality of technology teaching in compulsory school].

Chapter 5
Secondary Pupils' Perceptions of Their Experiences of Practical Work in Technology and Design

Kieran McGeown

Abstract This chapter investigates the perceptions held by pupils, across four Northern Ireland secondary schools, towards routine school-based practical work activities and project work in technology and design (T&D) and also that which had been contextualised within an industrial setting. The views of the heads of departments (HoDs) of each school and four local industrialists were also probed in relation to their views on the benefits or otherwise of both types of practical work.

The Questions I Asked and Why They Are Important

We live in a technologically advanced society within which the pace of technological change is rapid and where the effects of this rapid change affect virtually every facet of our lives. It is therefore vital that, when studying technology and design (T&D), our young people are aware of the key factors that affect the processes which ultimately bring about this technological change. Equally, it could be argued that they should also be made aware of the economic, social, ethical and environmental challenges that those involved in bringing about technological change must take into consideration.

A key aspect of pupils' experience of T&D in the UK is the 'practical work' that they undertake across the key stages (see Table 5.1). Irrespective of the perspective curriculum developers and teachers may hold in respect of the aims and content of technology education courses, practical activities are regarded as a principal component of the programme. The term practical work within a T&D scenario, as used in this study, refers to both the application of design ideas towards the manufacturing of a product or system that has an everyday use and activities which involve the use of hand tools or machines within a school-based manufacturing environment. A wide range of aims could define the nature of these activities, but with regard to the

K. McGeown (✉)
St Marys University College, Belfast, Northern Ireland, UK
e-mail: k.mcgeown@stmarys-belfast.ac.uk

P J. Williams, D. Barlex (eds.), *Explorations in Technology Education Research*, Contemporary Issues in Technology Education,
https://doi.org/10.1007/978-981-13-3010-0_5

Table 5.1 Key stages

Key stage (KS)	KS3	KS4
Ages of pupils	11–13 years old	14–15 years old
Type of school	Secondary	Secondary

Northern Ireland Curriculum (NIC) in general and the subject of T&D in particular, problem solving is seen both as a key thinking skill and a personal capability to be developed within pupils.

In a 2001 Department of Education for Northern Ireland (DENI) report, however, it was noted that only a small number of T&D departments in Northern Ireland (NI) had made links with local businesses but that these links had proved mutually beneficial. According to the report, particular benefits came in the form of raising the esteem of the teachers concerned and building up a team spirit between the schools and the participating businesses. There was however no mention of how these links to local industries had affected the learning experience of the pupils. This low level of links is notable given the stress put on developing pupils as 'contributors to the economy and the environment' as outlined within the statutory requirements of the Northern Ireland T&D Curriculum at key stage 3 (KS3) and also within the nonstatutory guidance which accompanied it.

The low uptake of T&D at key stage 4 (KS4) in NI and the small number of T&D departments identified as setting up links with industry in the 2001 DENI report provided both background and motivation for the questions that formed the nucleus of this study. It was further justified by the paucity of research relating to the subject, including the nature of the practical experience of pupils in NI schools, and by the absence of the pupils' voice in the studies that have been carried out. Indeed, as Benson and Lunt (2011) point out, there is a lack of pupils' voice in the literature more broadly. They further argue that 'Children naturally have very particular and important insights to offer in helping us to develop our understanding of teaching and learning' (p697).

The purpose of this study was twofold: to ascertain the views of both KS3 and KS4 pupils in relation to the practical activities they are currently undertaking and to ascertain whether or not they, and their teachers, think their schools having links to industry enriches the teaching and learning associated with problem solving through practical work in T&D.

Hence the two overarching questions that guided this research were:

1. What are the perceptions of secondary pupils regarding T&D with particular reference to their experience of their routine school-based practical activities and project work?
2. What are the perceptions of secondary pupils with regard to authentic, practical problem solving activities which have an industrial link?

It was hoped that answering these two questions would cast some light on the nature of secondary pupils' experience of routine school-based practical work and the perceptions of those young people towards practical work that has been contextualised within 'real-life' settings.

How I Tried to Answer the Questions

The main aim of this study was to try and ascertain the perceptions of a selected group of pupils studying T&D at secondary level in NI regarding their engagement with both routine school-based practical work and that which had a link to industry. This impacted considerably on the research design as, whilst every school provides a programme of routine practical work in T&D, it is much more difficult to identify those which offer pupils engagement with practical work that has links to industry. It was on this basis that it was decided that this investigation would focus on the experiences of pupils, from four NI secondary schools (see Table 5.2), who were involved in a European-funded project, titled 'Stimulating Science and Technology Competences Through Innovative Means for Teaching and Learning', STIMULA, (https://stimula.wordpress.com/), which had enabled both KS3 and KS4 pupils across the schools to undertake authentic problem solving practical work which had links to local industries.

In respect of their participation in the STIMULA programme, the Year10 groupings were involved in building a small-scale model of a hydro-electric turbine for the purposes of generating alternating current. The schools which the pupils attended had been partnered with a local NI-based industry which specialised in the manufacturing and installation of hydro-electric turbine units for small- and medium-sized businesses. The schools were then introduced by this firm to a local business to which it had supplied and installed a hydro-turbine system for the purposes of enabling it to be self-sufficient in terms of producing the electricity needed for both the business and the owners' domestic usage. The business in question was an organic farm which also had a restaurant for public use. The Year10 pupils from both schools subsequently visited the organic farm, talked to the owner about how the hydro-electric turbine was installed and were shown both how it was operated and maintained. The owner had been trained in all aspects of how to operate and maintain the hydro-turbine in order to maximise its output. Thus the pupils were informed how electricity was generated by the hydro-electric turbine and how this electricity was put to use in terms of making the farm self-sufficient in respect of electricity generation. The pupils were also made aware of how any excess electricity produced by the hydro-electric turbine was used in turn to supply the national grid, thus generating extra income for the farm owner. The owner was also able to explain the significance of the various geographical features of the farm that had affected both the design of the hydro-electric turbine system and methods used to successfully install it within the farm complex.

Table 5.2 Type of schools taking part in the study

School	Year group	Type of school	Age range of pupils attending school
School A	10 (KS3)	All boys junior high	11–16 years
School B	10 (KS3)	All girls grammar	11–19 years
School C	12 (KS4)	Co-ed grammar	14–19 years
School D	12 (KS4)	Co-ed grammar	11–19 years

The Year12 grouping within one of the co-educational grammar schools (14–19 years) was partnered with a local designer business. The pupils were taken through the design process that had been applied in the design, testing and manufacturing of a model that replicated a human throat. This model had not only to replicate the various parts of the human throat in terms of size and function but also the texture of human skin as the model had been commissioned by an American surgeon who needed it to test out state of the art laser controlled surgical incision techniques. The designer in this case not only facilitated the pupils visiting the business premises but also set specific design tasks for the pupils to undertake which he later planned to evaluate and provide feedback on to the pupils. This provided an opportunity for the pupils to work in two teams with the design brief of designing a device that would enable elderly people to transport their shopping from a car into their home with the minimum of stress. As in the other industrial visits, the owner of this local business was able to answer pupils' questions in relation to the opportunities for pursuing a career within this particular type of industry.

The Year12 group within the other co-educational grammar school (11–19 years) was afforded the opportunity to work in teams for the purpose of building and programming micro-robots for the purposes of accomplishing specific tasks. The pupils' work involved programming them to complete specific tasks in a particular order. The micro-robots were able to be programmed to move in a particular direction for a specified time and were also able to navigate around obstacles using light and infra-red sensors and bumpers which were effectively employed as switches. This particular school had been partnered with a local firm which manufactured and supplied apparatus for use in the field of medicine on a world-wide basis. The firm used robots as part of its production process, and the pupils were able to visit the firm in order to both witness how the robots performed their functions and to talk to the firm's engineer about how the robots were designed, programmed and maintained. The pupils were thus afforded the opportunity to contextualise their own work with the micro-robots with that of the industrial-based robots for the purposes of identifying similarities in terms of both the programming and the resulting functioning of the robots. The manager of the plant was also present to answer any questions that the pupils had in terms of possible careers within this particular industry.

The views of the pupils were sought in two stages. Firstly, their perception of their experiences of routine, school-based practical activities and project work was investigated. Secondly, after their participation in the STIMULA project, their perceptions were investigated as were those of their teachers and the local industrialists.

This study could have been carried out using a number of different data collection approaches, most obvious among which are the administration of questionnaires to pupils, the conducting of interviews and the observation and recording of practical work and industrial visits. Indeed, as permission had to be sought at the outset of the study, these possibilities were included in the information and consent documentation. Videoing was attempted at the beginning of the research however the schools were unable, due to timetable pressures, to allocate sufficient time for the process to be carried out effectively for the purposes of this research study.

Given these practical difficulties, it did not prove possible to adopt this approach to data gathering in the investigation.

As regards the use of questionnaires, it became apparent, through the literature review, that because there has been so little research about practical work in T&D, it was unclear what questions could be usefully asked. Whilst it would have been possible to ask very broad questions, in keeping with the exploratory nature of the investigation, it was considered that the young participants may find it burdensome to provide in-depth written answers to such questions. Seidman (2013) and Peraklya and Russuvuori (2011) argue that the purpose of interviewing is to understand the experience that people have in life along with the meaning that they take out of that experience. This is exactly the rationale that was applied to the selection of interviews as the primary data gathering tool within this study, i.e. the purpose was to find out the meaning that pupils made of their experience of engaging in both the normal school-based practical work and authentic problem solving tasks that had links with industry and, crucially, the issues that they wished to identify within these experiences.

A total of 44 pupils from 4 schools took part in Interview One and 41 in Interview Two connected with this study. Four HoDs and four industrialists also took part in Interview Two.

The pupils were interviewed on two separate occasions, whilst the industrialists and HoDs were interviewed on only one occasion after the pupils had both begun their industry-linked project and visited the industries concerned. Pupil Interview One was held during the months of October–November in 2012, whilst Pupil Interview Two was held during the months of February–April 2013.

In all cases the interviews with the pupils took place in their schools with the full permission and oversight of both the school principals and class teachers. The opportunity was taken in each school to meet up with the pupils taking part in the research as a group for the purposes of establishing a rapport with them by means of explaining both the purpose and the methodology of the research. During the interviews emphasis was placed on using the forenames of each pupil and putting them at ease before the interview started. Care was also taken to conduct the interviews in the morning when it was considered that the pupils would not be too tired and in the technology suite, a venue where they would feel as comfortable and as familiar with as possible. This also enabled the interviews to be concluded over the period of a morning without taking up too much time in the teaching day of the school whilst at the same time maximising the interaction time with the pupils. As indicated in all of the four schools taking part in the research, the pupils were interviewed in pairs except in one instance, in the Interview Two when, in School D, the teacher requested that one group be made up of three pupils. All interviews were recorded and transcribed for analysis.

Semi-structured interviews were also held with the HoD of each school. These were conducted during the months of June and July 2013 and were held at a time and place of the teachers' choosing. In the event, all were carried out in the school, either in a classroom or in the HoDs' offices. All were recorded and transcribed for analysis.

Finally, interviews were held with the local industrialists and experts who facilitated the pupils visiting local industries, again during the months of June and July 2013 and at the time and place of their choosing. Interviews with three of the industrialists were held in their place of work. The remaining interview with an industrialist was held via SKYPE at the request of the interviewee who would have otherwise been unavailable due to work commitments. Although not ideal in terms of conducting an interview per se, the SKYPE interview was successfully recorded and transcribed for analysis purposes. Details of the questions asked during the interviews with pupils, HoDs and industrialists are shown in Appendices 1, 2, 3 and 4.

The length of time taken on each of the above interviews did not exceed 30 min.

A thematic analysis of the pupils' interviews was carried out across the 2-year groupings, i.e. Year12s and Year10s; these groupings were made up of pupils across the four different schools. The first stage of immersion involved the compiling of any major themes emerging from a reading of all the interview transcripts. The second stage necessitated a closer reading of the transcripts for the purposes of identifying and suitably labelling as many categories as possible. This categoric analysis was carried out at the level of the questions. The third and fourth stages involved deleting or merging and relabelling of categories that overlapped. The fifth stage was concerned with rechecking the themes, stage six was a process of linking the themes and categories, and the seventh stage was the presentation of the overall findings.

Routine School-Based Practical Activities: A Summary of the Findings

What was most apparent in this study was the very positive response of the pupils to participating in practical activities and project work. This is an aspect of T&D which the majority of young people clearly enjoy and, indeed, many would have valued more opportunity to engage in such work throughout the programme. It was interesting to note that the pupils following the General Certificate of Secondary Education (GCSE) programme which offered a degree of choice were able to give more detailed overviews of their projects. It could be speculated that it may be the case that more choice fosters more engagement, but further research would be needed to explore this more fully. Typically, the activities had been used in the schools for a number of years. That having been said, there did appear to be a good balance between the projects offered to pupils, in terms of their focus, on either electronics or resistant materials. The Yr10 projects consisted of manufacturing a candle light holder and an electronic night light; both of these projects facilitated little design input by the pupils. There were a variety of Yr12 projects, all of which were electronic based; examples of these were a portable iPod docking station, a digital keypad lock and digital goal counter. All of the Yr12 projects facilitated some pupil input into the design of the finished prototypes.

Significantly, there did not appear to be an attempt at 'gendering' of the projects; in this particular instance, the pupils were all females.

Whilst practical project work was valued, some pupils did express a number of concerns about it. Some considered that the practical tasks were too quickly completed or that they were unable to complete them due to a lack of time or because their teacher decided that a particular task was too difficult for them to complete. The former suggests the projects may have been rather trivial. Some pupils indicated that they valued being able to design their own projects and that they would like to have a greater personal design input into the set projects. In similar vein, further reasons given by a few pupils for not valuing practical work as much as they considered they should were that the tasks were viewed as lacking in challenge, being 'childish' or being overly restricted to bench work. Some described their projects as being irrelevant to their everyday lives.

A number of those interviewed, especially the boys, reported that they did not value 'theory' for a range of reasons. A common perception was that there was 'too much theory'. The pupils suggested that there was no, or they failed to see any, relationship between the theory and the practical work they carried out. Overall, there appeared to be little appreciation of the role theory has in terms of developing technological competence in relation to specific projects. It was seen as separate from, rather than integral to, the technological activities they were undertaking. A number of pupils also associated theory with writing, often copying off the board. There was considered to be too much writing and, importantly, that it did not support their learning. Of particular significance to this study was the fact that, during the pupils' interviews, there was no explicit, specific mention made of problem solving nor, in the course of their discussions, of the key characteristics of problem solving processes whether in authentic contexts or not. For example, there was no reference to those problem-based scenarios which challenge pupils to discover knowledge and apply it as part of a solution and which literature suggests develops pupils' problem solving skills (Pedersen and Min 2003; Hill 1998; Hennessy and Mccormick 1994; Mettas and Constantionou 2007).

School-Based Practical Activities with an Industrial Connection: A Summary of the Findings

There was in general very positive feedback, among the younger and older pupils, to both the industrial visits and the school-based projects which had links to industry. However, there were some significant differences apparent in the learning experiences undergone by the pupils. For those pupils from Year10 (final year, KS3) whose project focussed on a hydro-turbine, it was evident from their responses that learning or consolidation of learning had occurred, as a result of their visit in a real-life setting, in respect of the use of water to generate electricity. Most expressed an understanding of how the use of a hydro-turbine for the generation of electricity brought both economic and environmental advances to society, such as an awareness of the social, ethical and economic aspects of technology. Which Dakers (2007) argues is necessary for pupils to develop technological literacy. A majority of the pupils also reported how they brought ideas gained during their industrial visit to bear on their school-based project with some mentioning that, through the project work, they had developed more generally applicable technological skills. One Year10 pupil commented ‘I always learn more like that, I’m not sure why. It’s like they know what they are talking about’.

The learning experiences undergone by the Year12 (final year KS4) pupils who visited a local industry which focussed on the design and ‘hands-on’ manufacture of products for various specialised markets and those who visited a local industry which focussed on the design and automated mass production of medical products highlighted some interesting differences. The former witnessed a process that involved the designing and manufacturing of a high-tech product for use in a hospital operating theatre. It was apparent that these pupils had a very enriching learning experience. A number observed that, as a result of the visit, they were more understanding of the fact that product design is an iterative process, involving modelling at its core, and where failure is regarded both as an integral part of the process and a normal occurrence. Pretzer et al. (2007) and Warner (2002) both contend that the fear of failure is an impediment to pupils being able to work creatively in D&T. Most of these pupils also referred to the fact that they became much more appreciative of how much the ethos and practice of teamwork was valued within such an environment and expressed an appreciation of how effective it was for them when they worked in teams for the project work set for them by the industrial designer. These pupils were also discussing, without any prompting, key aspects of problem solving as experienced in authentic situations. Significantly, and as noted above, this was not evident when other pupils had been discussing their experience of routine practical work. This may suggest that this type of learning environment reflects the criteria, as suggested by Hill (1998), Mettas and Constantionou (2007) and Hennessy and McCormick (1994), which enable authentic problem-based learning to take place. Most of these pupils indicated that they had learned how important teamwork was within this particular industry, whilst a majority suggested that their awareness of possible STEM-based careers had been increased as a result of the visit. One pupil suggested that it would be good for his school to have a strong relationship with local industry and that ‘if we spent time with them we’d maybe get to know more about it and possible career paths for us’.

The pupils who visited the industry specialising in the automated mass production of medical products did not, by contrast, appear to have had as deep an educational experience, especially in regard to the product design process. This highlights the fact that the nature of the industrial visit, as outlined by Post and Van der Molen (2014), should be matched to both the interests and learning needs of the pupils There appeared to be insufficient time, for example, given to allowing these pupils to talk to the engineers who controlled and programmed the workings of the industrial robots used in the industrial process that they had witnessed.

Overall, it is clear that the nature of the learning experience varied substantially depending on the industrial context in which it took place.

There was some interesting – and it is argued, significant – differences in the views given by the HoDs in relation to both the possible impact that they might have on pupils' learning and to the practicality of incorporating visits to industry into T&D programmes. Two HoDs in the schools under study were very positive in their review of the industrial link, and of having such links more generally, expressing little doubt about the potential that industrial visits have for providing an enriching educational experience for their pupils. Specifically, the HoDs of one of the schools that visited the hydro-turbine and the HoD of the school that visited the local industry which specialized in designing products for various markets contended that they believed that the industrial visits had been of considerable benefit to the young people.

In contrast, the HoD of the second school visiting the hydro-turbine was less positive, the suggestion being made that the pupils saw the industrial visit merely as a 'day out'. This was interesting, given that in fact the pupils from this school did provide evidence during the interview that suggested they had learned from their experience. The HoD of the school that visited the industry which specialised in the automated mass production of medical products also had reservations being unsure of the educational benefits to his pupils, specifically expressing doubt as to whether the pupils had been able to relate the production processes to what they were learning in T&D. In addition, he maintained that, due to the pressures of GCSE work, there was simply not enough time for industrial visits.

Overall, there was some evidence, drawn from both teachers and pupils, which would suggest that in at least some participating schools, there could have been a more focussed preparation for the industrial visits in respect of identifying the intended learning outcomes and sharing these with the young people. The teachers' perspective on and planning for the visit are important, in terms of extending the nature and degree of learning that occurs and a minimalist approach does little to enhance the learning experience of the pupils.

The industrialists all agreed on the potential for the enrichment of pupils' learning by means of schools-industrial partnership arguing that such visits offered opportunities for learning which was more reflective of authentic technological practice. Indeed, and in contrast, one described classroom-based practice as a 'form of learning that does not really relate, sometimes, to reality'. Another of the industrialists stated that enabling pupils to develop problem solving skills was hugely important and commented that 'For design it's crucial and it takes a bit of practice and independence'.

How This Might Be Used to Improve Teaching and Learning

This is, to the author's knowledge, the first study that relates specifically to the perceptions of Northern Ireland secondary pupils towards practical work in T&D. Given the steady decline of the number of pupils taking the Northern Ireland-based Council for the Curriculum, Examinations and Assessment (CCEA) T&D GCSE over the past 10 years, it was considered important to investigate this aspect of their courses due to the major contribution it makes to the study of T&D both as a subject within the revised NIC and at GCSE level. It is believed that this study adds to our understanding of how Northern Ireland pupils perceive practical work as presently undertaken in T&D at KS3 and KS4. Though no claim is made that the conclusions reached in this study can be generalised (it is accepted that they may relate only to those who participated), nonetheless, it is also believed that the findings offer important insights, indicative of issues that may be worth exploration in other countries and more widely in Northern Ireland.

With reference to the routine practical work, a majority of the pupils, and particularly the boys, indicated that they valued this aspect of their course and, indeed, that they would like to have more practical activities and project work included within T&D. However, some of the younger pupils also suggested that practical projects were too short, were too childish, were not challenging enough, were left uncompleted or were irrelevant to their and other's everyday lives. All of the aforementioned issues might, in part, be addressed by means of teachers determining the younger pupils' interests as part of the project design process and contextualising the project work within their everyday lives. The practice, as witnessed in most schools taking part in this study, of using the same projects over a number of years runs a considerable risk of making the subject matter both staid and uninteresting to pupils.

Some of the older pupils indicated that they did not value either the computer-based research work or the portfolios associated with GCSE projects. A number of pupils, particularly girls, indicated that they would have wished to have had a greater input into design decisions. Some pupils, but more boys than girls, also indicated that they did not value theory. Various reasons were given including a simple preference for practical work, the fact that it involved extensive copy writing, and also, importantly, because the young people saw little connection between the theoretical knowledge which they were studying and the projects they were undertaking. Theory therefore has to be both seen and understood by pupils as an integral part of the learning process and one which adds value to their overall teaching and learning experience when studying T&D. It is therefore important, some might even say necessary, that teachers ensure the theoretical knowledge being taught has direct relevance to the practical tasks being undertaken by the pupils.

It was somewhat surprising to see how few links that the pupils perceived there were between what they did in T&D and their everyday lives; indeed, there was not a single theme on which the majority of the pupils agreed. It was also notable that there was no mention of problem solving skills nor of the key characteristics of problem solving. It seems reasonable to suggest that both these factors might be linked to project work which is neither based on authentic problems nor considered as being sufficiently challenging by pupils. Again, and as suggested previously, teachers enabling pupils to become partners in the project design process might be one way in which these issues might be effectively addressed. Benson and Lunt (2011) argue that 'Children naturally have very particular and important insights to offer in helping us to develop our understanding of teaching and learning' (p697).

This study also gives an insight into pupils' perceptions as to how their experience of practical work might be affected by schools having a link with industry. It seems apparent from this study that there is potential for pupils to experience enhanced learning through their school having a link with an industry whose manufacturing processes provide the opportunities to contextualise their learning in relation to an authentic, real-world design and manufacturing environment. The obvious enthusiasm with which the majority of the younger and older pupils engaged with the practical work linked to industry does suggest this. Pupils indicated that as a result of the industrial linked project, they were more aware of STEM-based careers, had an increased awareness of the iterative nature of the design process, appreciated more fully the importance of teamwork, were more aware of how T&D can be used to meet a human need, and were better able to contextualise the associated project work in the real world. The study also provided some evidence to suggest, however, that the teachers could perhaps have prepared their pupils more fully for their visits.

The findings of this study have also had an impact on the author's professional practice. Working within initial teacher education has provided an opportunity to relate the pupils' perceptions to the professional development of prospective teachers of technology and design. It is therefore my intention to discuss the findings of this study, along with its strengths and the limitations, with my students so that they may consider it in relation to their future teaching of pupils in technology and design. Already, it has determined that I will put more emphasis on both differentiation and the importance of ensuring that practical work has relevance to the everyday lives of pupils. I also intend to emphasise that the practical work should be challenging, enabling the pupils to experience authentic problem solving. As part of this experience, it would be my intention to further facilitate the students in the development of their ability to teach theory effectively, so that the pupils can see a clear connection between it and the projects that they are undertaking. Pupils should also be encouraged to see 'failure' as an integral part of the design and manufacture process and not as an end in itself. I would also suggest to students and practising teachers that they consider the merits of establishing a link with local industry for the purposes of contextualising pupils' project work in the 'real world'.

Finally, and perhaps most importantly, I would advise teachers, based on the results of this research, to listen carefully to the voice of their pupils. If they do so, the evidence suggests that they will continue to develop in their role as professional educators.

The aim of this study was to investigate the perceptions of Northern Ireland secondary pupils towards practical work in T&D. It is vital that pupils experience practical work through the lens of effective problem solving in order that the aims of both the subject as outlined in the statutory requirements for T&D and of the revised NIC can be met. It is also believed that it has been demonstrated through both the words and actions of the pupils involved in this research that there is merit, with reference to enhancing the learning experience of pupils studying T&D, in schools pursuing an active partnership with local industries. It is recognized that this will involve new challenges for teachers in terms of both teaching strategies and the logistics in organizing such trips. It is also recognized that the nature of the learning experience will vary substantially depending on the industry visited.

It would appear therefore that, from what has been learned through this study, a reasoned argument could be put forward for conducting future research, with a larger, representative sample of pupils, to explore their perceptions of the amount of practical work undertaken and the time allocated for its completion, industry links, the appropriateness of its demand, the degree of choice and personal decision making, the role of theory, approaches to research, the relevance of T&D, and in particular practical projects, to their everyday lives, problem solving in T&D and links with other STEM subjects. This would also facilitate the exploration of any possible gender effects. Such research could suggest issues on which to focus in relation to the promotion of effective teaching of T&D.

It is also believed that, based on the evidence presented within this study, such partnerships have the potential to contribute to the development of an enriched learning environment within T&D which involves pupils engaging in authentic problem solving tasks which are grounded in real-world values.

Appendices

Appendix 1

Interview Schedule One

Question 1: What practical projects are you working on at the moment?
Question 2: What practical projects did you complete Years 8 and 9?
Question 3: What do you value doing in T&D and why?
Question 4: What do you not value doing in T&D and why?
Question 5: What type of projects and practical work would you like to be included in T&D?
Question 6: In what ways have the practical projects you have completed in T&D been useful to you in your everyday life?
Question 7: In what ways do you think that the practical projects you have completed in T&D might be useful to others?
Question 8: Which of the other STEM subjects do you think have links with T&D?

Appendix 2

Interview Schedule Two

Question 1: What did you learn from your visit to the local industry?
Question 2: Did you think that the visit to the local industry was a positive experience, and if so why?

Question 3: How useful to your everyday life was the industrial visit linked project?
Question 4: Did the visit to the local industry and associated project work enable you to better understand the links between STEM subjects, and if so, how?

Appendix 3

Interview Schedule for the Heads of Departments

Question 1: Did you think the visit to industry and associated project was a good idea? And if yes, why? And if no, why?
Question 2: Do you think the industrial visit enhanced the learning experiences of the pupils in relation to the associated project work? And if yes, why? And if no, why?
Question 3: Do you think that the industrial visit and associated project work was useful to the pupils in terms of being useful to their everyday lives?
Question 4: Do think there is merit in the school having a partnership with local industry? And if yes, why? And if no, why?

Appendix 4

Interview Schedule for Industrialists

Question 1: Do you consider that relating secondary school-based T&D project work to a real-life industrial environment brings value to this work? If yes, why? If no, why?
Question 2: Do you consider that visits to industry by post-primary pupils are of value to their everyday lives? If yes, why? If no, why?
Question 3: Do you consider that the visits to industry can enable pupils to better understand the connections between the STEM subjects? If yes, why?

References

Benson, C., & Lunt, J. (2011). We're creative on a Friday afternoon; investigating children's perceptions of their experience of design and technology in relation to creativity. *Journal of Science Education Technology, 20*(5), 679–687.

Dakers, J. (2007). Incorporating technological literacy into classroom practice. In M. J. De Vires, R. Custer, J. Dakers, & G. Martin (Eds.), *Analyzing best practices in technology design* (1st ed., pp. 125–137). Rotterdam: Sense Publishers.

Hennessy, S., & Mccormick, R. (1994). The general problem-solving approach in technology education. Myth or reality? In F. BANKS (Ed.), *Teaching technology* (1st ed., pp. 94–110). Buckingham: Open University.

Hill, A. M. (1998). Problem solving in real-life contexts: An alternative for design in technology education. *International Journal of Technology and Design Education, 8*(3), 203–220.

Mettas, A., & Constantionou, C. (2007). The technology fair: A project-based learning approach for enhancing problem solving skills and interest in design and technology education. *International Journal of Technology and Design Education, 18*(1), 79–100.

Pedersen, S., & Min, L. (2003). The transfer of problem-solving skills from a problem-based learning environment: The effect of modeling an expert's cognitive processes. *Journal of Research on Technology in Education, 35*(2), 303–320.

Peraklya, A., & Russuvuori, J. (2011). Analyzing talk and text. In N. Denzin & Y. S. Lincoln (Eds.), *The SAGE handbook of qualitative research* (4th ed., pp. 234–306). London: SAGE.

Post, T., & Van Der Molen, J. (2014). Effect of company visits on Dutch primary school children's attitudes towards technical professions. *International Journal of Technology and Design Education, 24*(4), 349–373.

Prezter, W., Rogers, G., & Jeffery, B. (2007). A model technology educator: Thomas A. Edison. *Technology Teacher, 67*(1), 27–31.

Seidman, I. (2013). *Interviewing as qualitative research: A guide for researchers in Education and Social Sciences* (4th ed.). Columbia University USA: Teachers College Press.

Warner, S. A. (2002). Teaching design: Taking the first steps. *The Technology Teacher, 62*(4), 7–10.

Part II
Stories of Technology

Chapter 6
Technology Landscapes in Children's Literature

Cecilia Axell

Abstract This chapter is about technology and children's literature. The aim of this study was to take a journey through the technology landscapes of a selection of Swedish children's literature, written during the last century (Axell, C. Technology landscapes in children's literature. A didactic journey from Nils Holgersson to Pettson and Findus. Dissertation. Linköping University, Linköping, 2015). The empirical material was based on catalogues used by school libraries to order new books, and some of the chosen books are still frequently borrowed from Swedish libraries. The selection was justified by the fact that they not only contain depictions of technology but also depict issues and problems considered relevant today in the field of technology education. Examples of these issues include reflections about the nature of technology, discussions about its advantages and disadvantages or the way technology is described in a social and historical context. The analysis of the stories showed that they not only contained depictions of technology but also depicted issues and problems relevant in the field of technology education. The ambivalent messages in the books revealed the multifaceted nature of technology as well as its complexity in ways that textbooks seldom do. A conclusion from the analysis was that children's fiction could be a starting point for creative discussions about the nature of technology and technology's effects on individuals, society and nature in the past, present and future.

The Questions I Asked and Why They Are Important

> The boy was greatly surprised to see all the sawmills that decked the shores. On Aln Island they stood, one next another, and on the mainland opposite were mill upon mill, lumber yard upon lumber yard. He counted forty, at least, but believed there were many more. "How wonderful it all looks from up here!" he marvelled. "So much life and activity I have

C. Axell (✉)
Department of Social and Welfare Studies, Linköping University, Norrköping, Sweden
e-mail: cecilia.axell@liu.se

P J. Williams, D. Barlex (eds.), *Explorations in Technology Education Research*, Contemporary Issues in Technology Education,
https://doi.org/10.1007/978-981-13-3010-0_6

not seen in any place save this on the whole trip. It is a great country that we have! Wherever I go, there is always something new for people to live upon." (Lagerlöf 1906/1908, p. 528)

Illustration by John Bauer (1882–1918) from *The Wonderful Adventures of Nils* (Lagerlöf 1906/1908)

In 1906, the children's book character Nils Holgersson flew over Sweden, and Swedish schoolchildren followed his adventures through Selma Lagerlöf's book *The Wonderful Adventures of Nils*. This was a Sweden in a state of change. The country was transforming from an agrarian to an industrial nation, and this industrialisation was dependent on the exploitation of iron ore, forests and hydro-electric power. Throughout his journey, Nils sees all sorts of technological innovations such as blast furnaces, lumber mills, steam engines, mowers and warships. These innovations arouse wonder in Nils about what humans can achieve. The book is fictional, but is also a geography textbook, written by one of the era's most prominent writers.

The Wonderful Adventures of Nils is not unique. Literature has long been used by educators as a pedagogical tool, and the importance of using children's books in elementary classrooms has been a topic of discussion in Sweden since the early twentieth century.

The research on which this chapter is based rests on a belief that children's literature can be considered a carrier of values and approaches to technology. Since fiction often illustrates how individuals experience social phenomena, it is also able to reflect our complex and contradictory nature better than some other genres (Greenberg and Schachterle 1992). Consequently, the aim of my study was to examine the messages conveyed about technology in a selection of Swedish children's books containing elements of technology and to explore how the views of technology relate to views of nature and views of the future. The study's overarching research question was:

> *What are the views of technology conveyed in children's fiction which contain aspects of technology, and how do these views relate to nature and future perspectives?*

My ambition was to contribute to an educational-inspired discussion of technology and children's literature, which also includes a historical perspective.

How I Tried to Answer the Questions

The analysed material consisted of Swedish children's literature written at different times, from the start of the twentieth century until today. I chose literature written over a period of time because I considered the historical perspective important in technology education. As a concept, technology is related to a time perspective in that it is associated with other concepts such as invention, development and change (Edgerton 2006).

Since there were no lists of literature that have been available in different school libraries over time, I chose to use catalogues used by school libraries to order new books. Some of the chosen books are still frequently borrowed from Swedish libraries. Initially, a large number of children's books (ca. 130 books) were examined, and children's books written by 6 Swedish authors were ultimately chosen: Selma Lagerlöf (1858–1940), Otto Witt (1875–1923), Elsa Beskow (1874–1953), Karl-Aage Schwartzkopf (1920–2009), Sven Wernström (1925–2018) and Sven Nordqvist (1946–). The selection was justified by the fact that they not only contain depictions of technology but also depict issues and problems considered relevant today in the field of technology education. Examples of these issues include reflections about the nature of technology, discussions about its advantages and disadvantages or the way technology is described in a social and historical context (Dakers 2006, 2011; de Vries 2005; Hallström and Gyberg 2011).

Technology landscape is a concept used when examining various environments through observing the different technological artefacts and the relationships between them. In other words, it is a kind of empirical "reality" that includes values and attitudes about technology (Hagberg 2009). This study could thus be described as a journey through the technology landscapes of some famous Swedish children's books, from the early 1900s to the early 2000s. For my theoretical and methodological starting points, I drew inspiration from a number of different research areas, mainly the fields of philosophy of technology, the history of technology and the history of ideas. As I used a hermeneutic method, my focus was to examine, understand and interpret the text at hand (Ricoeur 1981/2016; Ödman 2004). To synthesise what the texts had in common about technology, I used a comparative analysis which resulted in different themes.

In analysing the stories, I identified *what* types of technology are depicted, *how* they are depicted and *why*. I explored the *whys* by interpreting what the stories convey about the aims and purposes of technology and their assumptions about technology's effects on humankind, society and nature. Besides representations of actual technological items, I also considered linguistic symbols such as metaphors and analogies pertaining to technology.

To understand the central aspects of technology, I used Carl Mitcham's fourfold description of technology as *volition*, *knowledge*, *activity* and *object* (Mitcham 1994), combined with Thomas P. Hughes's systems perspective (Hughes 1987). These perspectives encapsulated the dimensions of technology central to my analysis. To describe the views of nature and futures conveyed, I used the notions of *anthropocentric* and *biocentric* views of nature (Sörlin 1991) and views of the future as *the effective society* or a utopian idea of *the good life* (Frängsmyr 1990).

Overall, the study had three analytical perspectives: the identification of *technologies* represented in children's books; the identification and interpretation of *the views of technology* transmitted in the stories, in relation to nature and future perspectives; and the identification of *shared and divergent themes* in the different stories.

What I Discovered

The results of my analysis showed that various facets of technology appeared throughout the 100-year period from which the children's books originated. Based on *what* kind of technology was represented in the stories, both Mitcham's (1994) aspects of technology and Hughes' systems perspective (Hughes 1987) were presented. With regard to Mitcham's description of technology as *volition*, *knowledge* and *activity*, these aspects' relationships to one another and their relationship to technology as *object* were clearly depicted.

Regarding *how* technology appeared in the chosen children's books, I identified six different themes:

- Technology as a *metaphor or analogy*
- Technology as *anthropomorphic*
- Technology as *autonomous*
- Technology as a result of a *creative driving force*
- *Masculine technology*
- Technology as *enduring*

Moreover, a clear duality appeared in the depiction of the technology: even if technology is fundamental to humankind, it also causes problems.

Technology as a Metaphor or Analogy

The metaphors and analogies in the stories could be described as being of two types: *explicit* (concrete) or *implicit* (abstract). Explicit metaphors are those that were developed from the natural world or the world of human beings and were used to elicit pictures of certain technologies and how they are used. When abstract metaphors were used in the stories, these stand for something greater, for instance, a phenomenon, event or social process. The abstract metaphors, however, were never explicit in the text, instead being present in hidden or obscured messages.

An example of an explicit metaphor from the story about Nils Holgersson is when Nils is taken captive by a family of bears and then introduced to the ironworks by Father Bear. Metaphors and analogies that are familiar to Nils and the reader are used to portray what happens in the ironworks:

> […] while the first bar of iron was being pressed, a second was taken from the furnace and placed under the rollers, and when this was a little along, a third was brought. Continuously fresh threads came crawling over the floor, like hissing snakes. The boy was dazzled by the iron. But he found it more splendid to watch the workmen who, dexterously and delicately, seized the glowing snakes with their tongs and forced them under the rollers. It seemed like play for them to handle the hissing iron. (Lagerlöf 1906/1908, pp. 366–367)

However, the depiction of the ironworks and the bears can also be interpreted as an implicit metaphor for the struggle between industrialisation/technological development (iron) and nature (the bear). Even though the bear claims that he and his ancestors have inhabited the region since the beginning of time, his conclusion is: "Let me say to you that iron is the thing that has given men the advantage over us bears, which is another reason for my wishing to put an end to the work here" (Lagerlöf 1906/1908, p. 368).

"Sandviken's Ironworks". Illustration from *The Wonderful Adventures of Nils* (Lagerlöf 1906/1908)

Implicit metaphors are also frequent in Otto Witt's *Technical Stories of the War for Young and Old* (1915). Several of the stories are expressions of what happens when humans use the destructive power of technology. However, the criticism expressed is not mainly concerned with technology itself, but more with human nature. For example, in the story "The Red Soldiers" red ants plunder their enemies and take slaves – even "toddlers". Parallels between ants and humans are drawn, and the anthill becomes an implicit metaphor for the ongoing World War I. When humans (represented by red ants) utilise their technological skills to develop destructive technology, they begin to destroy themselves. Similar to Mumford's (1967) description of the megamachine, the story depicts a hierarchically structured society in which individuals forget their basic needs.

Through the use of metaphors and analogies, associations of kinship are created between human beings, technology and nature. By linking the unfamiliar to familiar metaphors and analogies, understanding and acceptance for new technologies could be formed. In general, the explicit metaphors in the stories are more neutral in tone than the implicit ones. Criticism of society is often embedded in implicit metaphors, which also express a certain ambivalence about technology and technological development.

Technology as Anthropomorphic

Anthropomorphism – ascribing human traits and attributes, such as feelings, to animals or things – is quite common in folk tales and children's literature and also in the analysed books.

Illustration by Edwin and Ingeborg Lundborg from *Technical Stories for Young and Old* (Witt 1914)

In Witt's technical stories (1914, 1915), Schwartzkopf's books about the Choo-Choo family (1949/1966; 1950/1966) and Beskow's two fairy tales "Mr Klumpedump from Klumpedonien" (in *Storybook*, 1915/1954) and "The Green Car" (in *The Red Bus and the Green Car*, 1952), everyday gadgets come alive. In one of Witt's technical stories, even something more abstract comes alive; the weakest point of a railway bridge.

> The great railway bridge was dissatisfied; dissatisfied with itself, with the engineer who had designed it, with the workers who had built it, with the material of which it had been made, with the rivets, the stretch bars, the beams, the brackets … [T]his dissatisfaction had started one day when an unusually heavily loaded railroad train had passed over it. The weight had made the bridge tremble and then it felt something that can be compared to a human lumbago or something similar. There had been stretching and tension, and it had felt so strange in one particular place, in a joint near the centre of the bridge. And now the bridge was lying there pondering. It remembered all too well the words that the design engineer had often repeated: "A design is never stronger than its weakest point." (Witt 1914, pp. 92–93)

Technology is also highly humanised in the books about the Choo-Choo family. The artefacts have human feelings and needs and behave like humans. For example, trains give birth to "children" (train waggons), go to school and restaurants and do charity work. The humans, however, appear rather as enemies, as they constantly try to limit the trains' ability to act or punish them for not obeying orders. The threat to the train family is humans' need to evolve their technology, and there is therefore an imminent risk that they will abandon their older "servants" sooner or later.

Anthropomorphism helps to bridge the barrier between oneself and complex technology and contributes to building an emotional bond between human and machine (Schwarcz 1967; Waytz 2013, 2014). This intention is reinforced through showing how both children and machines can sometimes be disobedient and mischievous.

Technology as Autonomous

The cyborg Mulle the Tunnel Digger in one of Schwartzkopf's stories about the Choo-Choo family is an example of autonomous technology. Mulle is a hybrid between biological matter and machine but is nevertheless a result of human beings' technological activities (Haraway 1991; Wiener 1965):

> He looked like a mole, seen through a very large magnifying glass. Except for his fore paws, which were giant shovels, and hind paws, which consisted of large road scrapers. His nose was between his front shovels. Every time he snored, he sucked up dirt through his nostrils, which looked like a vacuum cleaner. The gravel then sprang directly out through his ears. (Schwartzkopf 1949/1966, p. 122)

As in Mary Shelley's novel *Frankenstein* (1818/1993), however, Mulle's creator loses control of his own invention, and if no one is able to stop Mulle, the world will be at risk from his constant digging.

In a similar way, in Beskow's fairy tale about "The Green Car" (1952), the technology (a car) is an anthropomorphic artefact with such a strong willingness it is self-driving. The new technology is difficult to control and scares both humans and animals. When the car finally ends up in the river, it is the older, more reliable technology (a farmer and his horses) which eventually carries home the self-driven car. The story can be interpreted as a metaphorical warning of what happens when humans lose control of their technology – it becomes unreliable and unpredictable.

The autonomous aspect of technology is also depicted in Beskow's fairy tale "Doctor Klokamundus' Invention" (in *Jollier: A Storybook*, 1919/1996). The story takes place in a fantasy land, Kringelkrokien, which is far ahead technologically. Inventions such as the telephone, the phonograph, the airplane and the cinema have existed there for several hundred years. But confidence in technology has gone too far, and technology is used as a tool to homogenise people. The children (boys) start to misbehave, but instead of seeing the underlying causes, the adults believe that the "problem" can be solved with technology: a high-tech fostering machine. Doctor Klokamundus, together with skilled engineers and builders, constructs the machine. The boys are locked into the machine, in which everything is managed by sophisticated and automated technology. If the boys long for someone to talk to, they can do so via a receiver on the wall, but the answers they get come from a gramophone. The story can be interpreted as an implicit warning against an over-reliance on technological solutions and when technology even replaces human coexistence.

The messages in these stories can be interpreted as modern technology following its own laws and thus being described as autonomous (Ellul 1964), but they can also be seen as warnings of the unpredictable consequences of technology (Winner 1977, 1986).

Technology as a Result of a Creative Driving Force

The stories about Pettson and his cat Findus (Sven Nordqvist) are centred on something that the protagonists invent and construct. By using existing technology, they create new technological solutions, for example, an alarm system in case the fox comes during the night, hungry and looking for a hen (Nordqvist 1986), a new cottage for Findus made out of the old privy house (Nordqvist 2012), a machine which is going to be a Santa Claus kind of robot (Nordqvist 1994) and a "fishing bow":

> Pettson had invented a fishing bow. Down by the lake he explained to Findus how it worked. The hook and the float were attached to an arrow. The arrow was attached to the fishing line. The rest of the line was wound onto a reel, which was attached to a bow. With it he could shoot the arrow with the hook far out into the water, much further than he could reach with the fishing rod. It worked quite superbly. (Nordqvist 1992, p. 11)

Pettson and Findus also use their artefacts in ways which are different from what they were originally intended for: an hourglass is rebuilt into an alarm clock, tweezers serves as sugar tongs, a hat is transformed into a lampshade and a small plane is used as a cheese slicer.

Don Ihde (2006) uses the term "the designer fallacy" when examining whether a designer can really include the purpose of an artefact in the design itself. Technology is embedded in different cultural contexts, which affect both design and applications. The same kind of technology could be used in different but specific contexts, and as an artefact technology seems to incorporate multiple uses and development trajectories, Ihde notes. In the same way, the books about Pettson and Findus show that the *context* governs the application of technology and the *user* determines its function. Daniel C. Dennett (1990) argues similarly to Ihde and concludes that it is not the inventor who ultimately decides what an artefact is or how it will be used – it is the user who decides.

Bricolage is a way of making use of and recycling old technology, or finding new ways to use existing technology (Lévi-Strauss 1966/1973). Since Pettson and his cat's inventions are virtually always constructed of components they already have, they can both be described as bricoleurs. By portraying Pettson and Findus as bricoleurs, the range of use for technology becomes greater. Moreover, as the stories about Pettson and Findus highlight, technological change or development is often about technology being transferred from one area to another (Edgerton 2006).

However, what distinguishes the books about Pettson and Findus from the other stories analysed is that many of their bricolages neither make things more effective nor solve a problem, and they often solve what, in a technological sense, can hardly be regarded as "problems". Examples include pots on wheels, tea cups with teaspoon holders, the Santa Clause machine and the alarm system to scare away the fox, which can more appropriately be described as artistic "Rube Goldberg machines" that perform something in a more complicated way than necessary

(Acharya and Sirinterlikci 2010). Pettson and Findus's technological solutions are the results of *creative activity* and are premised on a desire and a need to create. Pettson and Findus are thus both inventors and artists.

Masculine Technology

Technology is very often associated with boys and men, a notion confirmed in the analysed stories. In literary imageries of the prototypical inventor, this character is also practically always a man – from the technologically advanced heroes in Jules Verne's books to Gyro Gearloose, Professor Calculus and other ingenious figures in children's literature (Berner 2009). It is clear that the technology of Otto Witt's technical stories (1914, 1915), as well as in the story about Doctor Klokamundus' invention (Beskow 1919/1996), is something that belongs to male domains. The main characters (animals and artefacts) are mainly boys and males. One exception is a light bulb in one of the technical stories which is referred to as "she" (Witt 1914). The few women and girls present in the analysed stories have no interest in or knowledge about technology – they are mainly "supporting actors". Similarly, in Lagerlöf's story about Nils Holgersson:

> […] gradually he began to comprehend how much thinking and calculating men must have done before they discovered how to produce iron from ore, and he seemed to see sooty blacksmiths of old bending over the forge, pondering how they should properly handle it. Perhaps it was because they had thought so much about the iron that intelligence had been developed in mankind, until finally they became so advanced that they were able to build great works like these. (Lagerlöf 1906/1908, p. 369)

"Mine in Malmberget, Gellivare". Photo by "Ateljé Ekstrand". Illustration from *The Wonderful Adventures of Nils* (Lagerlöf 1906/1908)

The fact that technology is linked with male gender in these stories is not surprising. The technology concept at this time (the early 1900s) was associated with industrial and scientific development, and male inventors and scientists were held in high regard as heroes (Berner 1999). The "female-coded" technology and the areas where women dominated were not seen as being part of the technological sphere, and technology became something that belonged to the world of men (Oldenziel 1999). However, this explanation is not satisfactory, since children's books written almost a century later still mainly feature inventive male heroes. Pettson and his cat are a recent example of male, inventive protagonists. In relation to previous research, my study thus confirmed the fact that male engineers and inventors have been, and still are, portrayed as heroes and role models in children's literature (Hintz 2008; McCannon 2001). The stereotype of a handyman or an engineer is still a man (Berner 2009; Cockburn 2009; Mellström 2009).

Technology as Enduring

The technology landscape in Nordqvist's books about Pettson and Findus does not include any modern or state-of-the-art technology, but rather items such as kick-sleds, sledges, skis, radio sets, kitchen items, gardening tools and other artefacts that have a long history. What the books illustrate in this context is that even if new technologies emerge and other technologies disappear, there are also technologies that remain and continue to be used (Edgerton 2006; Kelly 2010).

In *A Rumpus in the Garden* (Nordqvist 1990), for example, various garden tools play an important part. The story begins on a spring morning when Pettson is standing in his vegetable patch with a spade in his right hand. With his other hand he is looking and feeling the soil:

> "Now's the right time," he said. "Today we can sow vegetables and set potatoes."
> Findus the cat ran around scaring beetles.
> "What do you mean SET?" he said.
> "Set in the ground. If we put carrot seeds down in the earth then carrots will grow there. And from each potato we set, there will be five to ten new potatoes." (Nordqvist 1990, p. 4)

The illustrations show how Pettson digs, uses his garden rake, sows his seeds and marks the furrows with flower sticks and empty seed bags. A wooden plough, a straw broom and a fly swatter are examples of other important artefacts in the story's illustrations. However, what turns out to be the challenge for Pettson and his cat is to create a technological solution that protects the vegetable garden from "intruders" (Pettson's hens, the neighbour's cows and a fox) without hurting any of them. Through underlining technology's enduring dimension, there is an appreciation for the technologies that have – historically and through the present time – been an important part of people's lives.

Different Views of Technology

With regard to technology and its outcome, the analysis showed that there are in general two sets of views in the stories. On the one hand, technology exists and develops independently of human and social forces (Ellul 1964). On the other hand, there is also a view of technology as a product of humans' social and cultural world (Mumford 1934/1963; Winner 1977). Based on these two overall views, the interpretation of *why* the technology is portrayed in specific ways in the stories resulted in four different themes or dimensions of views of technology:

- Technology as a *servant*: a view where technology is seen as a powerful tool serving to help humans achieve their dreams and aspirations.
- The *deterministic* view of technology: a view maintaining that humans have lost their domination over technology and technological advancement – technology constitutes a power of its own.
- The *nostalgic* view of technology: a view that early technology is better than modern technology and that old technology has a higher value.
- The *anti-consumeristic* view of technology: a view that campaigns for a sustainable technology society where objects are reused and technology is not made for economic profit.

In the stories, early technologies are generally given a higher value compared with industrial and state-of-the-art technology. This can be related to an idea that early technologies are in some way more democratic, since they are used in contexts where humans have a closer relationship with the technical construction and its use (Mumford 1964). For example, in some of the stories, crafts are contrasted with industrial mass production. Whereas crafts are produced by human hands, mass-produced objects have no "soul". This criticism of technology could be grounded in the belief that new and complex technologies are inevitably accompanied by negative consequences for nature as well as for human lives, human activity and the human mind.

The view of nature adopted in the majority of the stories could be described as *anthropocentric* or *weak anthropocentric*. The latter is a view of nature premised on the centrality of humans' needs but does not have a merely instrumental view of nature (Dobson 2000). Only one of the stories was interpreted as having a *biocentric* view of nature. Here, humans are pictured as a part of nature: the human's proper place is within the natural world, and humans and nature are seen equals.

By using Tore Frängsmyr's (1990) descriptions of two main views of the future, *the effective society* and *the good life*, it is clear that the effective society dominates the material, although a couple of books also depict *the good life*. However, my study diverges from Frängsmyr's on one important point. Frängsmyr posits that although belief in the power of technology seems contradictory to the utopia of the good life that respects nature's own way, the utopian idea actually also contains a hope for an effective society. The results of my study, however, showed that the analysed books representing the idea of a good life do not necessarily have any orientation towards future goals. In the stories, technology mainly solves workaday problems or even shows alternative solutions that may not necessarily be the most effective or the most modern ones. This result was also in contrast with previous studies that posit that children's fiction, specifically science fiction, as picturing dystopias (Applebaum 2006, 2010), since none of the examined books depict a dystopic view of the future.

In summary, my conclusion was that the presence of the technology in the books often conveys an ambiguous message: on the one hand, the technology is fascinating, essential and a result of creativity. On the other hand, technology is something that adversely affects both human relationships and nature.

How Children's Fiction Might Be Used to Improve Teaching and Learning in Technology

Technology in a Meaningful Context

As noted in the introduction to this chapter, children's literature has long been an educational tool, not only in literature teaching but also in other subjects. In the current Swedish curriculum for compulsory school (Lgr11), children's stories, fairy tales and myths take a central place in many of the subjects, from grades 1–6. The syllabus for crafts, for example, encourages the use of stories as sources of inspiration for the children's own creative work. For some reason, however, there is no mention in the curriculum that children's literature should be used in technology education. Nevertheless, the aim of technology education stipulates that:

> Through teaching, pupils should be given the opportunity to develop their understanding of the importance of technology and its impact on people, society and the environment. [...] Teaching should help pupils to develop their knowledge about the historical development of technology so that they are in a better position to understand complicated technological phenomena and contexts of today, and how technology has impacted and impacts on the development of society [...]. (Swedish National Agency for Education 2018, p. 292)

What the curriculum indicates is that today's world is an increasingly technological one, and in order to navigate their day-to-day life, students need technological skills in the form of highly developed technological literacy (Rohaan et al. 2010; Siu and Lam 2005; Turja et al. 2009). However, in a highly technological world, technology education cannot be expected to teach children how all modern technologies work. Hence, technological literacy implies more than the ability to create objects or to use or understand the function of certain technologies (Dakers 2011; Petrina 2000; Williams 2009). Since technology is created by humans, we must be able to judge it based on its consequences. Without this knowledge we will not be able to make deliberate democratic choices (Keirl 2006; Petrina 2000; Williams 2009).

The curriculum also highlights the importance of an historical perspective, on the grounds that historical knowledge will contribute to a better understanding of technological phenomena in a contemporary context. It can further our understanding about technology and human beings' changing relationship to it. By uncovering themes that help us understand our present time, a historical perspective can also benefit us in making well-reasoned decisions in the future regarding technology (Hallström and Gyberg 2011; Hallström 2013).

For many students, however, technology remains strongly associated with only artefacts, i.e. human-made items. This is not surprising since previous research has shown that there is an emphasis on artefacts and the making of artefacts in technology education. Likewise, students are commonly not given the opportunity to analyse the technology in a meaningful context. As a consequence, the connections between artefacts and humans, as well as what kind of implications the artefacts have in a societal context, are disregarded (Mawson 2010; Siu and Lam 2005; Turja et al. 2009). There are also studies showing that many students lack the ability to make critical judgments and rather act as uncritical consumers and users of technology (de Vries 2005). This artefact focus has also been confirmed by the Swedish Schools Inspectorate in its latest evaluation of technology education in Swedish compulsory schools, in which it is found that:

> Many of the schools under evaluation provided limited possibilities for pupils to develop the skills described in the syllabus for technology education. This is especially so with regard to developing reflection about technology and setting technology in a social and historical context. (Swedish Schools Inspectorate 2014, p. 27)

However, since knowledge in technology is often a matter of decisions and preferences, it also involves values that humans hold. Technology, in terms of making and using artefacts, is largely a practical activity which does not require much thought, but due to the inherent complexity and practical efficacy of modern technology, we also need to reflect on it (Mitcham 1994). Discussions and reflections should thus be regarded as an important part of teaching and learning about technology (Dakers 2006, 2011; de Vries 2005, 2006).

In the present study, I linked knowledge about technology to children's fiction and its educational function. As well as being a medium for mastering language, children's literature also performs a socialisation role, as it plays an important role in shaping how children think about and understand the world (Reynolds 2011; Zipes 1983). As socialisation is one aim of children's fiction, technological aspects have a legitimate place in the books, since technology is something that permeates much of human activity (Lee 1992; Schwarcz 1967).

As stated in the Swedish syllabus for technology, "[t]o understand the role of technology for the individual, society and the environment, the technology that surrounds us needs to be transparent and understandable" (Swedish National Agency for Education 2018, p. 292). That means, when technology is put into a broader perspective, like a fictional story, it can contribute to students' understanding of the technological world around them. By placing the technology within a meaningful context, teaching not only contributes to the understanding of technology but also makes the students see meaning in the teaching (Bjurulf 2011).

However, although the use of children's literature in technology classrooms seems to be common all over the world, technology in children's literature is still a neglected sphere of research within the field of technology education (Axell 2015, 2017b; Foster 2009). In science education, however, there seems to be a growing consensus among researchers that children's literature can be used to foster interest in, and positive attitudes towards, learning science in early years (see, e.g. Sackes et al. 2009; Monhardt and Monhardt 2006; Trundle and Troland 2005). Invitational use of children's literature may also offer a way for early childhood teachers to develop confidence and competence to teach science by inquiry (McLean et al. 2015). These research results can presumably be applied to technology education.

Using Children's Literature in the Technology Classroom

One suggestion is that the results of the present study can be used by technology teachers as a framework for analysis when planning to include children's literature as a pedagogical tool in technology education. The results may also serve as a basis for discussions in the technology classroom. By allowing books that interest students to be part of technology education, thoughts and ideas about technology can be presented as part of a context that engages them. However, it is important to keep in mind that children's literature is also a mediator of cultural values (Reynolds 2007; Zipes 1983). Choosing books in which technology is depicted in different historical and cultural contexts may therefore be essential, as they can contribute to an understanding of how human needs and problems can be technologically solved in different ways. The driving forces behind a technological change can be the same, but the solutions have been or remain different.

Examples of aspects that are worth keeping in mind when choosing a children's book to use in the technology classroom are:

- *What different aspects of technology could the story highlight?*
- *Does the story convey a wide or narrow picture of technology?*
- *How can the technological content of the story be reformulated to illustrate a "technological problem"?*

Children's literature is often described as having a pedagogical function because messages and values are often more explicitly expressed in literature for children, compared to books for adults (Reynolds 2011). However, this description is in contrast to the results of my study. In the analysed stories, the messages about technology and technological progress are ambiguous, and they are often characterised by an ambivalent view of technology and its consequences. On the one hand, technology is seen as something fascinating, exciting and meaningful for society. On the other hand, aspects of technology that hamper or harm human, society and nature are also present. The stories also seem to have a built-in duality, since they describe how technology can create individual freedom but also weaken human bonds.

Since fictional stories thus highlight the multifaceted nature of technology, they can encourage critical reflections about technology. Critical thinking is an important skill for students to develop in all subjects, not least in the subject of technology (see, e.g. Axell 2017a; Keirl 2017; McLaren 2017; Petrina 2000). Asking questions that invite reflection can therefore be regarded as an important part of teaching and learning in technology. As fictional stories often contain different types of technology, teachers can support students' thinking and reflections by asking open-ended questions. These kinds of questions stimulate the children's curiosity and desire to discuss, as they invite a closer examination of a certain "phenomenon". Good questions do not have a given answer, but encourage students to discover that there are several ways of looking at the same thing. Examples of questions could be:

- *What kinds of technologies are represented in the story?* A suggestion is to let the students categorise the technologies found on the basis of what kind of human needs or problems the specific technologies are created to fulfil or solve.
- *How is technology depicted in the story*? An identification of how the technology is depicted in a story can bring an interest and curiosity to students and their own relationship with the technology they are using and are surrounded by. For example, the *anthropomorphic* and *autonomous* aspects of technology can open up a discussion about artificial intelligence and other technological solutions that are about to enter our lives, such as self-driving cars and robots in elderly care. Moreover, by assuming that our social *gender roles* are predominantly shaped by ideas, students can develop an awareness of how norms, gender stereotypes and gender patterns are created in relation to technology.
- *What view of technology does the story convey?* Is it a view that technology serves human wants and needs (a *servant* view), a view that technology "lives" its own "life" independently of humans (a *deterministic* view), a view that older technology is better or more reliable than new technology (a *nostalgic* view) or a view that technology should be sustainable and not only be used or created for economic profit (an *anti-consumeristic* view)? And *who is in charge of technological development?* As humans, we are responsible and obliged to take into account everything in our environment, including the natural environment. An identification of the views of technology and views of nature in the stories can highlight an important aspect: the relationship between humans, technology, nature and sustainable development.

By asking questions, teachers can challenge students' perceptions and ideas and support the development of critical thinking skills. Thinking critically is about drawing conclusions, evaluating and seeing things from different angles, i.e. opening up and considering alternative ways of seeing. Based on fictional stories, it is possible to make comparisons between different kinds of technological solutions as well as investigating how humans have created technology throughout history to fulfil their needs or to solve problems. The stories may thus encourage discussions about technology's impact on individuals, society, prevailing norms, gender, nature and the planet as a whole (Axell 2017a, b; Axell et al. 2014).

A fictional story can also give an opportunity to introduce specific technological terms and concepts, as well as serving as a starting point for technological activities.

If a children's book serves as a basis for a practical activity, it is an advantage if the story includes a plot that can be identified as a "technological problem".

My conclusion, after a careful analysis of a number of children's books, is that the technology landscapes in children's literature could contribute to broadening and expanding technology education. The ambivalent messages in many stories reveal technology's multifaceted nature and its complexity and thus make it possible to problematise different aspects of technology. Fictional stories can thus serve as a platform for open-ended enquiries and dialogues about its nature and driving forces behind technological change, as well as the effects of technology on individuals, society and nature in past and present times. Fictional stories can also be connected to practical activities in technology and prevent technology education from becoming unreflected "doing" activities. In this way, fictional stories can function as a pedagogical tool to fulfil the overall objectives of the Compulsory School Technology Curriculum.

Additional Considerations

Finally, an additional important aspect for teachers to bear in mind is that cultural norms include societal norms about gender (Crisp and Hiller 2011; Gooden and Gooden 2001; Oskamp et al. 1996; Trepanier-Street and Romatowski 1999). Under-representation of girls and women and gender stereotyping in children's picture books have been documented in previous research (Hamilton et al. 2006; Martin and Siry 2009; Paynter 2011). In combination with the fact that technology itself is often considered to belong to the male domain, this makes it important to be careful in the choice of literature. For example, books categorised as "facts for youngsters" often include content with a heavy focus on males and masculine-coded technology (Axell and Boström 2015). One can assume that there is an imminent risk that children's literature may contribute to socialising children into traditional roles and limit their interest in other activities that may also suit them (Gooden and Gooden 2001; Oskamp et al. 1996). This means that even if children's fiction can serve as a basis for an introduction to reflections about technology, there is also a risk that stories will contribute to preserving stereotyped gender patterns. When choosing children's literature, the technology teacher hence needs to consider two perspectives: technology *and* gender.

Teachers can actively challenge "hidden" messages by opening up conversations, asking questions and supporting activities that make the students start reflecting about how the relationship between gender and technology is presented in the stories:

- *What characteristics can be used to describe the character who is engaged in technology-related activities in the story? Are those typically "boy" or "girl" traits? Why?*
- *Would the story be different if the main character were the opposite gender? If so, how? Would they make the same choices in the story? Why or why not?*

- *How can the technological content of the story be reformulated to illustrate a "technological problem" which is not masculine- or female-coded?*

The students can be given the task of rewriting the story with characters of the opposite gender or rewriting the story with gender-neutral characters by using names that are given to both men and women and avoiding the pronouns "he" and "she".

Students can also be encouraged to write their own "technology stories" with a reverse/or neutral gender perspective and where technological protagonists develop solutions that are in harmony with nature as opposed to dominating nature.

To sum up, the fact that the characters in children's literature who use or create technology – whether they be human, animal, a tractor or a train – are almost always male can either be seen as a problem when choosing literature or be an important issue in discussions with students.

The Analysed Children's Books

Beskow, E. (1915/1954). *Sagobok* [Storybook]. (5th ed.) Stockholm: Wahlström & Widstrand.

Beskow, E. (1919/1996). *Muntergök: Sagobok* [Jollier: A storybook]. (New ed.). Stockholm: Bonnier Carlsen.

Beskow, E. (1952). *Röda bussen och gröna bilen: Bilderbok till Johan från farmor* [The red bus and the green car]. Stockholm: Bonnier.

Lagerlöf, S. (1906/1908). *Nils Holgerssons underbara resa genom Sverige. Bd 1–2* [The Wonderful Adventures of Nils, Part 1–2]. Stockholm: Bonnier.

Nordqvist, S. (1986). *Rävjakten* [The fox hunt]. Bromma: Opal.

Nordqvist, S. (1990). *Kackel i grönsakslandet* [Rumpus in the garden]. Bromma: Opal.

Nordqvist, S. (1992). *Pettson tältar* [Pettson goes camping]. Bromma: Opal.

Nordqvist, S. (1994). *Tomtemaskinen* [The Santa Claus machine] Stockholm: Opal.

Nordqvist, S. (2012). *Findus flyttar ut* [Findus moves out]. Stockholm: Opal.

Schwartzkopf, K-A. (1949/1966). *Familjen Tuff-Tuff* [The Choo-Choo Family]. (3rd ed.). Stockholm: Geber.

Schwartzkopf, K-A. (1950/1966). *Hemma hos familjen Tuff-Tuff* [At home with the Family Choo-Choo] (2nd ed.). Stockholm: Geber.

Witt, O. (1914). *Tekniska sagor för stora och små* [Technical stories for young and old]. Stockholm: Hökerberg,

Witt, O. (1915). *Krigets tekniska sagor för stora och små* [Technical stories of the War for young and old]. Stockholm: Hökerberg.

References

Acharya, S., & Sirinterlikci, A. (2010). Introducing engineering design through an intelligent Rube Goldberg implementation. *Journal of Technology Studies, 36*(2), 63–72.

Applebaum, N. (2006). The myth of the innocent child: Interplay between nature, humanity and technology in contemporary children's science fiction. *The Journal of Children's Literature Studies, 3*(2), 1–17.

Applebaum, N. (2010). *Representations of technology in science fiction for young people*. New York: Routledge.
Axell, C. (2015). *Technology landscapes in children's literature. A didactic journey from Nils Holgersson to Pettson and Findus*. Dissertation. Linköping: Linköping University.
Axell, C. (2017a). Critiquing literature: Children's literature as a learning tool for critical awareness. In P. J. Williams & K. Stables (Eds.), *Critique in Design and Technology Education* (pp. 237–254). (Contemporary Issues in Technology Education). Singapore: Springer.
Axell, C. (2017b). Technology and children's literature. In M. de Vries (Ed.), *Handbook of technology education* (pp. 1–17). Cham: Springer International Publishing.
Axell, C., & Boström, J. (2015, April 6). Facts for youngsters – Contextualised technology or fragmented artefacts? A study on portrayals of technology in picture books from a gender perspective. In M. Chatoney (Ed.), *PATT 29 plurality and complementarity of approaches in design and technology education* (pp. 42–48). Marseille: Aix Marseille University.
Axell, C., Hallström, J., & Hagberg, J.-E. (2014). Images of technology and sustainable development in Swedish children's literature. *Australasian Journal of Technology Education, 1*(*1*), 1–9.
Berner, B. (1999). *Perpetuum mobile? Teknikens utmaningar och historiens gång*. Lund: Arkiv.
Berner, B. (2009). Teknikens kön. In P. Gyberg & J. Hallström (Eds.), *Världens gång – teknikens utveckling: Om samspelet mellan teknik, människa och samhälle* (pp. 279–293). Lund: Studentlitteratur.
Bjurulf, V. (2011). *Teknikdidaktik*. Stockholm: Norstedt.
Cockburn, C. (2009). On the machinery of dominance: Women, men, and technical know-how. *Women's Studies Quarterly, 37*(1/2), 269–273.
Crisp, T., & Hiller, B. (2011). Is this a boy or a girl?': Rethinking sex-role representation in Caldecott Medal-Winning Picture books, 1938–2011. *Children's Literature in Education, 42*(3), 196–212.
Dakers, J. R. (2006). Towards a philosophy *for* technology education. In J. R. Dakers (Ed.), *Defining technological literacy: Towards an epistemological framework* (pp. 145–158). New York: Palgrave Macmillan.
Dakers, J. R. (2011). The rise of technological literacy in primary education. In C. Benson & J. Lunt (Eds.), *International handbook of primary technology education: Reviewing the past twenty years* (pp. 181–193). Rotterdam: Sense.
Dennett, D. C. (1990). The interpretation of texts, people and other artifacts. *Philosophy and phenomenological research, 50*(Supplement), 177–194.
Dobson, A. (2000). *Green political thought* (3rd ed.). London: Routledge.
Edgerton, D. (2006). *The shock of the old: Technology and global history since 1900*. London: Profile Books.
Ellul, J. (1964). *The technological society*. New York: Vintage Books.
Foster, P. N. (2009). An analysis of children's literature featured in the "Books to Briefs" column of technology and children, 1998–2008. *Journal of Technology Education, 21*(1), 25–43.
Frängsmyr, T. (1990). *Framsteg eller förfall: Framtidsbilder och utopier i västerländsk tanketradition*. Stockholm: Allmänna Förlaget.
Gooden, A. M., & Gooden, M. A. (2001). Gender representation in notable children's picture books: 1995–1999. *Sex Roles, 45*(1–2), 89–101.
Greenberg, M. L., & Schachterle, L. (1992). Introduction: Literature and technology. In M. L. Greenberg & L. Schachterle (Eds.), *Literature and technology* (pp. 13–24). Bethlehem: Lehigh University Press.
Hagberg, J.-E. (2009). Att lära i teknikens rum och landskap. En metadidaktisk betraktelse. In P. Gyberg & J. Hallström (Eds.), *Världens gång – teknikens utveckling: Om samspelet mellan teknik, människa och samhälle* (pp. 41–60). Lund: Studentlitteratur.
Hallström, J. (2013). Teknikhistoria öppnar upp vidare perspektiv på tekniken. In J. Hallström & C. Klasander (Eds.), *Ginners teknikdidaktiska handbok: Några teser om teknik, skola och samhälle* (pp. 61–72). Linköping: Linköping University Press.

Hallström, J., & Gyberg, P. (2011). Technology in the rear-view mirror: How to better incorporate the history of technology into technology education. *International Journal of Design and Technology Education, 21*(1), 3–17.
Hamilton, M. C., Anderson, D., Broaddus, M., & Young, K. (2006). Gender stereotyping and under-representation of female characters in 200 popular children's picture books: A twenty-first century update. *Sex Roles, 55*(11–12), 757–765.
Haraway, D. J. (1991). *Simians, cyborgs and women: The reinvention of nature*. London: Free Associations Books.
Hintz, E. S. (2008). Heroes of the laboratory and the workshop: Inventions and technology in books for children, 1850–1950. In M. M. Elbert (Ed.), *Enterprising youth: Social values and acculturation in nineteenth-century American children's literature* (pp. 197–211). New York: Routledge.
Hughes, T. P. (1987). The evolution of large technological systems. In W. E. Bijker, T. P. Hughes, & T. J. Pinch (Eds.), *The social construction of technological systems: New directions in the sociology and history of technology* (pp. 51–82). Cambridge, MA: MIT Press.
Ihde, D. (2006). The designer fallacy and technological imagination. In J. R. Dakers (Ed.), *Defining technological literacy: Towards an epistemological framework* (pp. 55–59). NewYork: Palgrave Macmillan.
Keirl, S. (2006). Ethical technological literacy as democratic curriculum keystone. In J. R. Dakers (Ed.), *Defining technological literacy: Towards an epistemological framework* (pp. 81–102). New York: Palgrave Macmillan.
Keirl, S. (2017). Critiquing as design and technology curriculum journey: History, theory, politics and potential. In P. J. Williams & K. Stables (Eds.), *Critique in design and technology education* (pp. 109–133). (Contemporary Issues in Technology Education). Singapore: Springer Nature.
Kelly, K. (2010). *What technology wants*. New York: Viking.
Lee, J. Y. (1992). The feminization of technology: Mechanical characters in picture books. In M. L. Greenberg & L. Schachterle (Eds.), *Literature and technology* (pp. 206–224). Bethlehem: Lehigh University Press.
Lévi-Strauss, C. (1966/1973). *The savage mind*. London: University of Chicago Press.
Martin, S. N., & Siry, C. A. (2009). Raising critical issues in the analysis of gender and science in children's literature. *Culture Studies of Science Education, 4*(4), 951–960.
Mawson, B. (2010). Children's developing understanding of technology. *International Journal of Technology and Design Education, 20*(1), 1–13.
McCannon, J. (2001). Technological and scientific utopias in Soviet children's literature, 1921–1932. *Journal of Popular Culture, 34*(4), 153–169.
McLaren, S. V. (2017). Critiquing teaching: Developing critique through critical reflections and reflexive practice. In J. P. Williams & K. Stables (Eds.), *Critique in design and technology education* (pp. 173–192). (Contemporary Issues in Technology Education). Singapore: Springer.
McLean, K., Jones, M., & Schaper, C. (2015). Children's literature as an invitation to science inquiry in early childhood education. *Australasian Journal of Early Childhood, 40*(4), 49–56.
Mellström, U. (2009). Män i teknikens värld. In P. Gyberg & J. Hallström (Eds.), *Världens gång – teknikens utveckling: Om samspelet mellan teknik, människa och samhälle* (pp. 295–310). Lund: Studentlitteratur.
Mitcham, C. (1994). *Thinking through technology: The path between engineering and philosophy*. Chicago: University of Chicago Press.
Monhardt, L., & Monhardt, R. (2006). Creating a context for the learning of science process skills through picture books. *Early Childhood Education Journal, 34*(1), 67–71.
Mumford, L. (1934/1963). *Technics and civilization*, 1. (Harbinger books ed.). New York: Harcourt, Brace & World.
Mumford, L. (1964). Authoritarian and democratic technics. *Technology and Culture, 5*(1), 1–8.

Mumford, L. (1967). The myth of the machine. In *Technics and human development* (Vol. 1). New York: Harcourt Brace Jovanovich.

Ödman, P.-J. (2004). Hermeneutik och forskningspraktik. In B. Gustavsson (Ed.), *Kunskapande metoder inom samhällsvetenskapen* (pp. 71–93). Lund: Studentlitteratur.

Oldenziel, R. (1999). *Making technology masculine: Men, women, and modern machines in America, 1870–1945*. Amsterdam: Amsterdam University Press.

Oskamp, S., Kaufman, K., & Wolterbeek, L. A. (1996). Gender role portrayals in preschool picture books. *Journal of Social Behavior and Personality, 11*(5), 27–39.

Paynter, K. C. (2011). *Gender stereotypes and representation of female characters in children's picture books*. Published dissertation. Liberty University.

Petrina, S. (2000). The politics of technological literacy. *International Journal of Technology and Design Education, 10*(2), 181–206.

Reynolds, K. (2007). *Radical children's literature: Future visions and aesthetic transformations in juvenile fiction*. New York: Palgrave Macmillan.

Reynolds, K. (2011). Introduction. In K. Reynolds & M. O. Grenby (Eds.), *Children's literature studies: A research handbook* (pp. 1–10). Basingstoke: Palgrave Macmillan.

Ricœur, P. (1981/2016). *Hermeneutics and the human sciences: essays on language, action, and interpretation, Cambridge philosophy classics edition*. New York: Cambridge University Press.

Rohaan, E., Taconis, R., & Jochems, W. (2010). Reviewing the relations between teachers' knowledge and pupils' attitude in the field of primary technology education. *International Journal of Technology & Design Education, 20*(1), 15–26.

Sackes, M., Trundle, K. C., & Flevares, L. M. (2009). Using children's literature to teach standard-based science concepts in early years. *Early Childhood Education Journal, 36*(5), 415–422.

Schwarcz, H. J. (1967). Machine animism in modern children's literature. *The Library Quarterly, 37*(1), 78–95.

Shelley, M. W. (1818/1993). *Frankenstein, or, the modern Prometheus*. Ware: Wordsworth.

Siu, K. W. M., & Lam, M. S. (2005). Early childhood technology education: A sociocultural perspective. *Early Childhood Education Journal, 32*(6), 353–358.

Sörlin, S. (1991). *Naturkontraktet: Om naturumgängets idéhistoria*. Stockholm: Carlsson.

Swedish National Agency for Education (2018). *Curriculum for the compulsory school system, the pre-school class and the leisure-time centre 2011*. Revised 2018. Stockholm: Skolverket.

Swedish Schools Inspectorate Report (2014:04). *Teknik – gör det osynliga synligt. Om kvaliteten i grundskolans teknikundervisning*, 2014. http://www.skolinspektionen.se/Documents/Kvalitetsgranskning/teknik/kvalgr-teknik-slutrapport.pdf

Trepanier-Street, M. L., & Romatowski, J. A. (1999). The influence of children's literature on gender role perceptions: A reexamination. *Early Childhood Education Journal, 26*(3), 155–159.

Trundle, K. C., & Troland, T. H. (2005). The moon in children's literature. *Science and Children, 43*(2), 40–43.

Turja, L., Endepohls-Ulpe, M., & Chatoney, M. (2009). A conceptual framework for developing the curriculum and delivery of technology education in early childhood. *International Journal of Design and Technology Education, 19*(4), 353–336.

Vries, d. M. (2005). *Teaching about technology: an introduction to the philosophy of technology for non-philosophers*. Dordrecht: Springer.

Vries, d. M. (2006). Technological knowledge and artifacts: An analytical view. In J. R. Dakers (Ed.), *Defining technological literacy: Towards an epistemological framework* (pp. 17–30). New York: Palgrave Macmillan.

Waytz, A. (2013). Making meaning by seeing human. In K. D. Markman, T. Proulx, & M. Lindberg (Eds.), *The psychology of meaning*. Washington, DC: American Psychological Association.

Waytz, A. (2014). The mind in the machine: anthropomorphism increases trust in an autonomous vehicle. *Journal of Experimental Social Psychology, 52*(3), 113–117.

Wiener, N. (1965). *Cybernetics or control and communication in the animal and the machine*. Cambridge, MA: M.I.T. Press.

Williams, J. P. (2009). Technological literacy: A multliteracies approach for democracy. *International Journal of Technology and Design Education, 19*(3), 237–254.

Winner, L. (1977). *Autonomous technology: technics-out-of-control as a theme in political thought.* Cambridge, MA: MIT Press.

Winner, L. (1986). *The whale and the reactor: A search of limits in an age of high technology.* Chicago: The University of Chicago Press.

Zipes, J. D. (1983). *Fairy tales and the art of subversion: The classical genre for children and the process of civilization*. London: Heinemann.

Chapter 7
Using Narratives for Communicating the Nature of Technology

Sachin Datt

Abstract Humans have entertained themselves with stories since the beginning of civilization. Researchers have attributed many reasons for humans indulging in this seemingly leisurely activity. One of the reasons is transferring cultural knowledge through generations. As humanity moved across different sets of cultural values, its narratives changed accordingly. The narratives of religious cultural values are characteristically distinct from narratives of cultural values that represent modernity and so are narratives expressing the culture of science. What are the characteristic features of narratives that carry the cultural values of technology? How can these narratives help in communicating the nature of technology? These are some of the questions we explore by analysing technology narratives from the history of airplanes.

The Questions I Asked and Why They Are Important

The question I asked is subset of the question 'Do narratives have some kind of underlying structure?' If there are, then what are they?

This question has been answered in the past literature. I am further extending this question to ask 'what are the structures underlying technological narratives?' What can we do with the knowledge of the structure of a technological narrative?

The questions related to understanding the nature of narrative are based on the assumption that though there are an infinite number of narratives in the world, they have common, repetitive elements that allow them to be grouped under clusters of similar properties. Such a work was undertaken by Vladimir Propp (Propp 1968) who discovered common, repetitive narrative elements in Russian folk tales. The limitation is that knowledge of structural qualities of Russian folk tales can help a writer create or identify another Russian folk tale, but not any other type of tale, or

S. Datt (✉)
Sushant School of Design, Ansal University, Gurgaon, India
e-mail: sachindatt@ansaluniversity.edu.in

P J. Williams, D. Barlex (eds.), *Explorations in Technology Education Research*, Contemporary Issues in Technology Education,
https://doi.org/10.1007/978-981-13-3010-0_7

any tale that is completely unique. However, there are some scenarios where writing a particular category of tale may be a necessary requirement. For example, if one has to write a story about superman, then it has to have certain elements that make it a superman story. It will be distinct from a batman story, even though both will have similar elements that make them part of the larger category of a superhero story. Knowledge of what constitutes a superhero story is essential to be able to write or judge a superhero story. Even to appreciate a break in the tradition of a superhero story, the audience must have some existing expectations from a superhero story; otherwise, they will not even notice the break in convention. These exceptions are built up by repetitive elements appearing again and again in subsequent stories.

Technological narratives, being distinct from scientific or any other type of narrative, can be used in technology education contexts to communicate the nature of technology to students. The narratives provide a rough idea of what to expect in a life lived for introducing some technological advancement in a domain. This has relevance especially in the Indian context where we see a society that enjoys the technological advancement of the modern world, but a survey of Indian narratives shows world-views still embedded in ancient mythology. A particularly popular Indian narrative portrays that the first successful flight in the world was made by an Indian. With the knowledge of technological narratives, we can identify that the claims made about an Indian making the first successful flight cannot be true because its narrative structure does not even fall under the category of a technological narrative. With the help of technological narrative structure, we can have some idea of what kind of events to expect in a technological narrative.

How I Tried to Answer the Questions: Method

To find answers to these questions, we analysed autobiographical narratives of the Wright brothers (Wright 2008) and Otto and Gustav Lilienthal (Lilienthal 1896), the people largely responsible for flying and controlling the first heavier-than-air machines. Our objective was to find out common and uncommon story events in their narratives. We also studied the biographies of George Cayley and Langley (White 1914), but the extensive detail of the experience of flying was only available in the Wright brothers' and Otto and Gustav Lilienthal's description of problems encountered during attempts to fly. Some of the features of their narratives had similar events. The analysis was performed to observe the instances when the exploration of the protagonist experienced an impeding force and when the protagonist was able to move forwards and what was the intensity of these impeding and enabling forces.

For this purpose, I looked at some sections of the historical narratives of technology and compared it with the general structure of a narrative with beginning, middle and end. Looking at a narrative with its beginning, middle and end gave an idea of elements to be searched for in the narrative. I have not discovered any new elements of a narrative. I have only tried to find those 'general' narrative elements that are present ‘particularly’ in a technological story. For example, the beginning part of any narrative generally consists of initial situation in the life of protagonists. The initial situation in a story starts with a place and some protagonists, for example, a school and a boy studying in it or a farm and a farmer with his family. In the case of technology narrative, we found that the initial situation of course is about some place and a protagonist, but it is particularly about a protagonist being inspired by some existing innovation introduced to them during their routine life. In the case of the Wright brothers, it was the toy helicopter which their father purchased for them in their childhood.

Similarly, we found many common elements that constituted the middle and end part of a technological narrative. In this study, we are sharing those ‘general’, beginning, middle and end part of a technological narrative, which we found in historical narratives of airplanes.

What I Found Out

Here, some of the common elements of the stories in the history of flight are outlined. They are not ordered in the same sequential manner as in the original story. The various points can occur in different sequences depending on the style of the narrative presentation.

1. The narratives of technological innovation suggest that the lead protagonist or a group of protagonists get introduced to some **‘existing’ technological** artefact that arouses their curiosity.
2. The lead protagonists gain information about the person who is associated with the existing technology and get inspired by their work.
3. The means by which the lead actors get information about the people associated with existing technology could be by book, public lecture, media or word of mouth.
4. The lead protagonist begins to imitate the existing technology in some secret manner in a place that secures the protagonist or the group’s privacy. It is known to only few people.
5. The place has some resources to start building their initial prototypes. The resources are accessible to the protagonist directly or are procured through effort.
6. The protagonists get involved in the existing technology as a sport. The word ‘sport’ has been used explicitly in the writings of the Wright brothers and Otto and Gustav Lilienthal. Elsewhere in the history of electricity, Joseph Priestley describes working with electrical apparatus for the purpose of ‘entertainment of friends’. The technological explorations are done with the spirit of entertainment.
7. How the protagonists find a secure place to begin their work could have another story behind it. It could be unique for all individual stories. It could even be an established institute that provides necessary resources.

8. The presence of a helping mentor who provides material, technical and emotional support has been seen repeatedly in narratives of flight. In the case of the Wright brothers, it was Mr. Chanute, who himself had performed many flying experiments.
9. There is a tension between the protagonists' everyday life which may include family or survival pressure, a regular job and the 'technological sport' that the protagonists are pursuing. The tension may be weak or strong in individual cases. But the fact that the protagonist has to find some way to take out time from their regular life and devote it to their sport is a significant part of the narrative. It requires conscious decision by the protagonist to continue or quit the sport under everyday life's pressures.
10. It is his decision to step out of the everyday life that differentiates people who **explore their passion for some time and forget it** and those who **exercise the sport to more and more challenging levels.** It is not just the everyday life that invites the person towards itself, but the sport itself becomes mundane and repetitive if the protagonist does not do more to conquer some other greater challenge in the sport. At that juncture, there is a temptation to quit the sport and return to the everyday practical life. The sport in itself does not provide a sustained and lifelong passion. The sport that enthused earlier can quickly lead to boredom within weeks or even days. It is through the protagonists' own curiosity and hunger for further exploration, and the company of other people exploring the same sport, that keeps the interest in the sport alive. This point is understood clearly through the life of the Wright brothers. Their first attempt in childhood was to construct their own toy helicopters. They made many numbers of them. The technology for constructing a toy helicopter existed at that time. Many children would be content with constructing toy helicopters, but the Wright brothers tried to do more. As they grew up, they tried to construct heavier-than-air machines of greater and greater sizes. They imitated existing design, but they also added a little bit of their own exploration in the existing model. But as they increased the size of the model, the designs, which worked perfectly at small dimensions, failed miserably in larger size.
11. The initial failure encountered, while trying to achieve a new level of problem or challenge, can be detrimental for further progress. Many people would believe that there is no point expanding the boundary of the technology that already exists. What is the chance that we will find out something new?
12. The drive to emerge from lack of self-belief can only be overcome if the protagonist has **prior experience of unknown problem-solving**. In the case of the Wright brothers, luckily, their experience of manufacturing bicycles provided them enough self-confidence that many technical problems can be resolved, if one tries to identify the problem and work over it. A person who has not had prior experience of problem-solving in some other simpler technology is likely to quit at this point. There is no prior experience of success to fall back to at this point to recover from failure. Otto Lilienthal was an engineer by training. It may be purely a matter of luck that some people get the experience of solving real-world problems from a young age by being born in a certain

environment. Once this initial failure is overcome and the protagonist continues to solve one problem after the other, of various degrees, then there are greater chances of the protagonist introducing a new technology.

13. As the protagonist advances further in exploration of the sport, they may encounter some distinct core technological problems that were not identified previously. They may be required to give up reliance on existing knowledge. Up till this point, the existing knowledge helps the protagonist, but from this point on, the existing knowledge itself becomes a hindrance for the person to move forwards. Here, the protagonists have to give up their reliance on existing data and create their own data from their own unique observation and apparatus.
14. The nature of this core problem has been described well by Altshuller, the creator of the theory of inventive problem-solving. He says that creativity in technology emerges when a conflicting situation is resolved. The conflict is of the kind when a solution which worked perfectly in some previous situation itself becomes the cause of the problem in a different situation (Horowitz 1999). This conflict can be seen very clearly in the Wright brothers' attempt to create a flying machine. The first core problem that the Wright brothers faced was regarding the problem of balancing a glider in air. This problem was partially overcome previously by Otto and Gustav Lilienthal in their many successful engineless flights. However, their glider failed when the wind speed was more than a certain limit. The machine was simply uncontrollable in turbulent wind conditions. Otto Lilienthal even lost his life due to this shortcoming of their glider. The problem was in the basic scientific principle on which the previous glider design relied. In the previous design, the glider acted as a pendulum (Figs. 7.1a and 7.1b).
15. As soon as it tilted in one direction, it automatically corrected itself by returning back to the initial position. This self-correcting mechanism was a huge advantage in engineless glider flights. It keeps the flight in a state of balance. But when the wind speed exceeds a certain limit, this pendulum movement of the glider itself becomes the reason for the pilot to lose complete control over the flight making the glider turn over and crash. The Wright brothers identified this as the core problem of existing flight and gave up building the glider cantered on the science of the pendulum movement. The gliders they made did not have this self-correcting mechanism. It was completely controlled by the

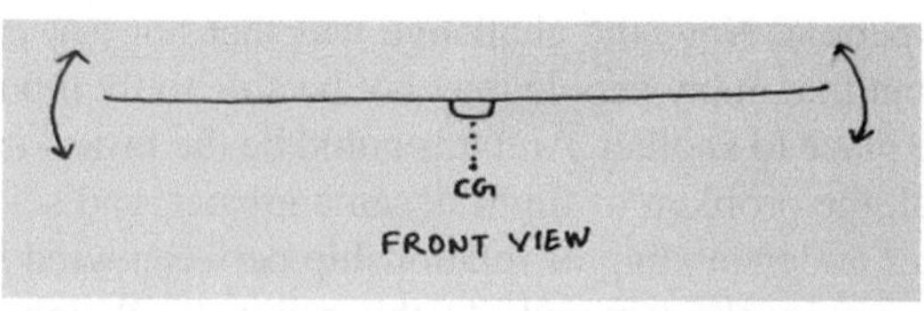

Fig. 7.1a If the CG is below axis of rotation, the wings swing like a pendulum, returning to the stable position automatically. (**b**) The profile of the wing can also be imagined as a pendulum; the difference here is that the pendulum is constantly pushed upwards by air pressure which is balanced by extra weight on the front part

Fig. 7.1b The profile of the wing can also be imagined as a pendulum; the difference here is that the pendulum is constantly pushed upwards by air pressure which is balanced by extra weight on the front part

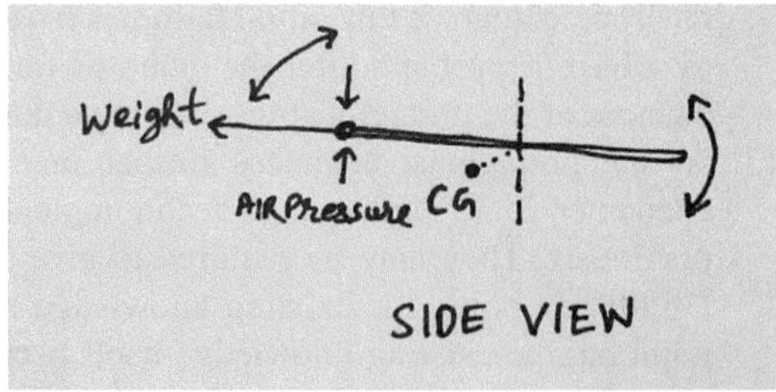

pilot who controlled the movement of all the various types of wings attached. And this change in reliance upon some traditional belief made all the difference.

16. At two critical points in pursuit of this technological sport, science enters into the picture. The first point is when the protagonist looks at the sport not merely as his or her own source of enjoyment but when they decide to make this sport accessible to all people. The moment goal of the sport is expanded from mere pleasure for one or two people to a source of enjoyment for many, the problems and challenges visible to the innovators vastly increase which they could not see earlier. The life of Otto and Gustav Lilienthal is a testimony to this fact that they took great pleasure in practising the sport of flying a glider. However, they could have been content once they made some rudimentary flights in the air which they successfully did. But their goal was not just that the two of them should fly. They specifically wanted to master this sport to such an extent that any human should be able to experience the joy of flying just as they experienced it. But for that to happen, they required greater knowledge of the behaviour of a glider as it interacted with all kinds of wind conditions. Without science, gaining this knowledge was impossible. Hence, Otto and Gustav Lilienthal created their observation tables after every flight and came to some strong beliefs about the nature of interaction between wind and wing. Science brought the knowledge of flight to such an extent that someone other than Otto and Gustav Lilienthal could make their own glider. This knowledge helped the Wright brothers (next generation) to start early in their youth.
17. The second stage when science entered the picture of flying machine technology was when the Wright brothers saw flying as more than only a sport. For long, it was a sport for the brothers just as it was for Otto and Gustav Lilienthal. But at some point, the Wright brothers saw that a flying machine could be a means of transport for people. Now, the challenge was that not just anybody should be able to fly it but that many people may not be able to fly it but still could travel in it from one place to another. And this could be the fastest means of transport. With this goal, the problem of flight became greater, and science was required to have greater understanding of relationship between wind speed, weight and lift than was previously required. At this point the Wright brothers realized another set of anomalies in existing data about relationship between air pressure, weight and lift. With the increased weight, due to propeller motor, the lift generated was wrongly predicted according to the existing mathematical tables

created by Langley. The Wright brothers again did their own unique experiments to create their own tables of pressure, weight and lift relationship, which was more accurate than the data that was previously available.

18. Apart from the core problems that require intervention of science, there are other minor unpredictable problems that keep sprouting up till the final successful technology attains complete shape. These minor problems have no general character. They could be unique to each individual story.
19. One of the common problems is the search for vendors who specialize in making some specific part of a machine. The innovators have to perform extensive survey of the market and know their vendors well. But in the case of the Wright brothers, dissatisfaction from existing vendors compelled them to design their own petrol engine. Still, the search for a vendor had to be performed even to come to the conclusion that no vendor could provide the required machine part.
20. Once satisfied, the protagonist prepares a public demonstration. Initially, the protagonist may be overenthusiastic by the grand future possibilities that the success of a given technology may bring about. It is here that the temptation to come out of the workplace where the technology took birth compels the protagonists to come out into the world and display their achievement to the larger public. The lure of fame, accolades and recognition that a successful trial may bring could be destroyed in an instant by failing in front of public and media.
21. The protagonist has to return back to the workplace and overcome more unique problems that could not be foreseen before.
22. Once resolved, the protagonists decide to demonstrate the new possibility to a close group of people who really matter rather than to display the innovation to the whole world.
23. The protagonist may achieve some recognition from unexpected channels in the form of some funds that they receive for developing the technology further.
24. Once the technology has been recognized secretly, without media attention, the protagonists are busy communicating with prospective patrons or any interested party (it could be someone from the media also but not representing the mainstream).
25. The technology slides into the world and into the lives of people without much media recognition, and that is its success. Media recognition may come much later.

With these elements of a technological story, one can find gaps in a story that is centred on invention or improvement of some technology. If we compare this narrative structure with the film Hawaizaada (2015) about the Indian who invented the world's first flying machine, we can see gaps in the narrative of the Indian story. In the story, the protagonist takes inspiration from ancient Indian scriptures written in Vedic period (4000 BC) to select the fuel for the engine, as mercury. But there is no precedence of any technology that associates mechanical work done by a mercury-fuelled engine by any other inventor. As we have seen in the technological narrative elements described above, **no invention comes about in a vacuum. It is based upon some existing successful innovation.** There are many such gaps that can be found in the Indian story, and the story can be declared as invalid or deceiving.

The technological elements described in this research may help in writing technological narratives and also help in checking the validity of a 'technology story'. As more stories are analysed, we can find further corroboration for the presence of certain narrative elements, or we may also find instances where some of these narrative instances are not present. It may also induce us to ask questions when we are presented with a technological narrative. Which existing technology is the new technology based on? What are the successful elements that are incorporated from the existing technology? When does the success of the existing technology reach its limit? What new elements are added to the existing technology to expand its boundary of limitation? How did the protagonist get interested in that specific domain of technology? What was their inspiration? Which was the most difficult problem they encountered? How and when did they receive initial recognition from peers? Who were their mentors? Who were the core group of people acquainted with the work in the beginning?

How This Research Can Be Used to Improve Teaching and Learning

The Problem of Motivation or Motivation in the Wrong Direction

Carroll and Gill (2017) has reported that there has been a decline in the uptake of design and technology subject from 98% to 28% by students enrolled in GCSE programmes in England (country). In the context of India, design and technology education has still not been able to make inroads into school curriculum. The skills of building things have been added as part of science education in the National Curriculum Framework 2005 (NCF 2005). But in the existing scheme of things, building something is simply seen as an extension of a craft-like activity where students build some physical model. Most of the models that students build, as evidenced from the various science fairs, are replicas of models available on YouTube, from remote control cars to lifting arm robots, etc. There is no data available on this topic about what kind of models students are actually building in schools as part of their science curriculum. It is based on the author's own experience of conducting and visiting many science fairs that schools conduct in schools of Delhi and Mumbai. This is an even more undesirable situation because schools believe that through such science fairs, the need of design and technology curriculum in schools is already being fulfilled, and hence there is no need for any dedicated subject. Though there is motivation for such science fairs, they are aimed at mere exhibition of craft skills or explaining some scientific principle like laws of motion, laws of conservation of mass, etc. through demonstration rather than sharing of problem-solving experiences encountered during the process of creating a 'new' artefact that is different from something already existing.

Overall Curriculum Planning: Integration of Storytelling in Project-Based Learning

We propose to integrate storytelling with project-based learning. In project-based learning, students are expected to work on a problem, do research and come up with a solution by creation of some form of artefact. The journey of any creator is a story of ups and downs. Each creator goes through a series of small successes and failures before arriving at the suitable solution if they choose to complete the whole journey. Stories of inventions show the journey of famous inventors and their internal motivation that provides the energy for pursuing a new path. For example, students working on designing gliders can immediately connect their hurdles and problems with those faced by original creators like Otto and Gustav Lilienthal or the Wright brothers. They will see similarity and differences in their problems and those faced by earlier inventors.

First, the students have to be given opportunity to work on a project so that they gain sufficient experience of the difficulties that they may face. Having spent time working on a specific task and not able to find a solution, the ground is prepared to tell a story where a similar problem is addressed by some inventor in history. The greatest difficulty in keeping the momentum of 'a building and making task' is not being able to find a way out of some problem. It is very easy to give up and get distracted by some other work like solving imaginary mathematics problems for which experience is already gained to feel good about oneself or go off to some other entertaining task and forget about the problem itself. A story at this point can provide interesting associations that can keep the focus of the student in mind even when they are doing any other relaxed activity in evening. They can come back remotivated to work on the unsolved problem again.

The story has to start not directly by evoking some historical event but from the context emerging from the classroom itself. The teacher can take the students themselves as the main characters in the story and bring in and connect the local story with the story of historical inventors. The main characters who have names of the students in class arrive at a point where they get motivated to create something. But they face some realistic hindrances that the students actually face in the project. At the point when they are struggling to find solutions, they are reminded of the story of some historical inventors. For example, for the project of making a glider, it is clear that their initial designs will be failures, and the feeling will soon set in that it is impossible to design a glider. At that point the story of the Wright brothers or George Cayley can be told and principles of flight explained through the story of George Cayley's discovery.

But it has been seen through several workshops that telling a story does not result in desired changes in students' motivational approach towards the project. The real impact is when students are asked to tell their story of going through the project. It has been observed informally through many workshops that when students were told that at the end of their project they have to tell the story of their positive and negative experiences after the project, their level of participation and engagement is visibly higher than when they are only asked to see the project as a problem-solving exercise. The stories that students tell not only give information about their project but also can make the teacher know a lot about the student's life experiences. This is because students recall the names of other people who helped them made the project and in which way. The students if they want can take photographs of them working on the project and tell their story with pictures at the end of the project.

But there is one more thing here. From our informal experience of attempting to implement this research, we found that the teachers' interest in the story and her own attempts of creating something are the core driving force. If the teacher is not motivated by the story she is telling, then this approach does not work. The teacher cannot mechanically select any story about invention and use it in a design class and expect results. The teacher should have discovered first something for her own self in the historical story before using it in the session plan.

Here is a link to students in action attempting to make 'new' paper gliders that actually fly: https://www.youtube.com/watch?v=B0gqDE5P_60.

Assessment for Learning

After each storytelling session, students can be asked to recreate another prototype/solution to the problem. It is difficult to assess the individual motivation level of students, but one can definitely observe how many students in a class create the second prototype that was better than their previous prototype. It is generally observed that when students are asked to repeat the whole process and make a second prototype after dumping the first prototype altogether, a good number of students make a prototype worse than the previous one. Some students do not even turn up to the workshop to do rework when their first attempt is rejected. Only a handful are seen redoing the work all over again. One can assess the change in the quality of the second, third and fourth prototype for majority of students after subsequent storytelling (both historical and students' own narrative) and reworking sessions. The change in the quality of subsequent prototypes can be an indicator for the positive effect of the story.

Qualitative assessment is also possible by analysing the subtle experience that students share. The stubbornness of the material world throws up unknown surprises at the creator. A student who has actually attempted to solve multiple problems will always have some unique problems to share. The problem may not at all be related to the core design of the product but some other related issue. For example, 'when I was trying to carry this model from home to school, it broke apart in transportation, so I had to figure out a way to package it so it could sustain the hammering of (Indian) roads'.

This is not an exact science. More work needs to be done on assessing the effect of storytelling session on students and the stories that students tell.

Conclusion

Using stories for starting a teaching or learning session in class has already gained wide acceptance in education. Many teachers make use of story resources to give a motivating start to their lesson plan. Stories particularly in the context of design and technology education can enable students to appreciate the contribution of inventors/innovators to technology as part of their understanding of the nature of technology. This can help students develop a perspective on technology about how technology works and avoid alienation from our technology-based society. However, one needs to be careful in selecting the right kind of stories for design and technology classes. Not all stories give the correct idea of the essence of events that lead to an innovation. Some stories give genius-like qualities to inventors who come up with ideas in a flash of intuition without giving details of the past inventions that the inventor has referred to. Such a problem is very prevalent in the Indian context. Inventions are shown to get created out of some mythical superpower of the genius inventor. The rough structure of narratives of technology that we have presented in this research can help teachers in choosing and selecting narratives that provide the essence of technological innovation events that maybe closer to reality. The assumption being that acquaintance with real technological narratives can ultimately inspire students to look deeper for solutions, rework prototypes and continue to remain motivated to solve a problem for a longer amount of time and collect sufficient experiences to tell their story to people around them.

References

Carrell, M & Gill, T. (2017). Uptake of GCSE subjects 2016. Statistics Report Series No. 114. Cambridge, UK: Cambridge Assessment.

Hawaizaada. (2015). *Hawaizaada*. In Wikipedia. Retrieved March 12, 2017. https://en.wikipedia.org/wiki/Hawaizaada.

Horowitz, R. (1999). *Creative problem solving in engineering design*. Retrieved from Advanced systematic inventive thinking: http://asit.info/Creative%20Problem%20Solving%20in%20Engineering%20Design,%20thesis%20by%20Roni%20Horowitz.pdf

Lilienthal, O. e. (1896). Practical experiments for the development of human flight. *The Aeronautical Annual* (pp. 7–20). Retrieved from invention, psychology, msstate: http://invention.psychology.msstate.edu/inventors/i/Lilienthal/library/Lilienthal_Practical_Exp.html

NCF (National Curriculum Framework). (2005). *National Council of Educational Research and Training*. New Delhi.

Propp, V. (1968). *Morphology of the folktale*. Austin: University of Texas press.

White, G. (1914). *The Aeroplane: Romance of reality series*. London: T.C. & E.C. Jack.
Wright, O. &. (2008, May 11). *The early history of the airplane.* Retrieved from drugfreereading: http://www.drugfreereading.com/interest_novels/Early%20Airplane%20HistoryHIST.html

Part III
Planning and Pedagogy

Chapter 8
Teachers' Assessment Practices

Eva Hartell

Abstract This chapter is based on my doctoral thesis *Navigare necesse est*—Necessities and complexities regarding teachers' assessment practices—which is grounded in my prior experience as a teacher of elementary science, technology, engineering and mathematics (STEM) subjects. It focuses on teachers' assessment practices in primary and lower secondary technology education and contributes to the fields of technology education and educational assessment. In this chapter, I summarize my findings and describe methods of providing affordance for teachers' assessment practices to bridge teaching and learning in technology education.

The Questions I Asked and Why I Think They Are Important

This chapter is based upon a thesis that aims to deepen and generate knowledge regarding teachers' assessment practices. Assessment is a broad area because teachers use it in many ways to assess pupils' knowledge. As a teacher, I used various approaches to develop my assessment practice, some of which were beneficial while others were not. When given the opportunity to conduct research, I decided to develop teaching and learning through four sub-studies about primary and lower secondary technology classroom practices in Sweden. I first explored teachers' formal documenting practices, minute-by-minute classroom assessment, their self-efficacy on assessment and their statements and motives relating to criteria for success while assessing students' e-portfolios. I combined this with a literature review to investigate, describe and exemplify compulsory school teachers' assessment practices. This quest stems from my own teaching experience and the following observation by Lindström: 'A teacher who fails to assess what the students do, cannot decide whether or not she is contributing to or impeding their progress'

E. Hartell (✉)
KTH Royal Institute of Technology, Stockholm, Sweden
e-mail: ehartell@kth.se

P J. Williams, D. Barlex (eds.), *Explorations in Technology Education Research*, Contemporary Issues in Technology Education,
https://doi.org/10.1007/978-981-13-3010-0_8

(Lindström 2006). The questions I asked during my journey towards a PhD stem from that experience.

The first question I investigated was the usability of the Swedish mandatory assessment document *Individual Development Plan with written assessments* (IDP) and how teachers utilize it in their follow-up and instruction planning for supporting student learning in technology education. The IDP is regulated by the Swedish Education Act (NAE 2009). As a STEM teacher, I completed an IDP for every student twice a year for the subjects I taught, namely, technology education, physics, biology, chemistry and mathematics. This was hard work, not only in the effort involved but also in terms of describing in writing where the pupils were in their learning journey compared to national standards and suggesting possible ways to progress.

The rather depressing results from the first study and the existing educational literature highlighted the importance of short-cycle formative assessment and encouraged me to develop the second sub-study which explored how teachers conduct minute-by-minute follow-up in their classroom.

During 2009–2015, when I was working on the thesis, Sweden undertook several education reforms. Sweden has a teacher-based assessment system, and some subjects, not including technology, have national tests to support teachers in grading. Owing to the changes in the Swedish Education Act of 2011 (*Skollagen Svensk författningssamling 2010:800* 2013), only those teachers with a teacher certificate were allowed to award grades in Sweden. However, the majority of technology instructors in Swedish schools lack training in technology subjects, general teaching, or both. Hence, the research question of our third sub-study revolved around whether subject-specific teacher education influenced teachers' self-efficacy beliefs as related to assessment in technology. We conducted a quantitative study using a questionnaire to investigate the answers to this question.

The fourth sub-study evolved from the first three and aimed to investigate the factors that influenced teachers' decision-making during classroom assessment. Specifically, we sought answers to the criteria for success that primary teachers emphasized during assessment.

So in summary, the research questions were:

(RQ1): how is the IDP-mandatory assessment document used by primary school teachers in their follow-up and future planning of their student's knowledge development in technology?

RQ2: how is teachers' minute-by-minute follow-up enacted in classroom?

RQ3 considers factors regarding teachers' self-efficacy beliefs regarding their own views on assessment practices in technology.

RQ4: what criteria for success do primary teachers emphasize during the act of assessment?

How I Tried to Answer the Questions

The sub-studies employed both quantitative data collected from official governmental databases, software-generated statistical data from questionnaires and qualitative methods such as observations, analyses and interviews.

The first sub-study examined IDP documents from five municipalities. We investigated classroom practice by observing two teachers, Karen and Karl, while they taught a sequence of technology lessons in school Years 4 and 5 (10-/11-year-old pupils). Both teachers chose and validated the technology topics and the instructional methodology for the classroom observations in order to minimize disturbances to student progress and the teaching schedule.

The second sub-study was undertaken in two teachers' classrooms. In Karen's class on mechanics, students made jumping jacks, built cars and visited the fire station to look at real-life applications of mechanisms. Karl's pupils explored the history of communication, including the Morse and other alphabets; written communications such as runes, cuneiform, etc.; and electric circuits and safety. The five key strategies for formative assessment were (1) clarifying and sharing learning intentions and criteria for success, (2) engineering effective classroom discussions and other learning tasks to elicit evidence of pupil understanding, (3) providing feedback that moves learners forward, (4) activating pupils as instructional resources for one another and (5) activating pupils as the owners of their own learning (Black and Wiliam 1998) that was used as tools for analysing the data collected.

The third sub-study investigated the teachers' self-efficacy on assessment in technology. A 5-point Likert scale survey was used instead of interviews. We received 651 responses (response rate of 96.5%) from teachers working in schools participating in a school development program funded by the European Social Fund, in Stockholm region. Informants who claimed they taught technology (88) were then divided into two groups, firstly those school staff working as technology teachers with academic credits in technology education and secondly those who did not have academic credits in technology and teacher education. Data from the survey and also national statistics from official databases was analysed with quantitative measures.

The fourth sub-study involved six teachers and a class of 11-year-old pupils. The pupils designed a model robot to complete various tasks at home, such as recording NHL games, walking the dog, completing homework, scanning and submitting homework and baking cupcakes. During classroom activities, the students built Web-based synchronous e-portfolios of their learning and the product development using text, photos, moving pictures and sketches on their iPads. Teachers assessed these portfolios through adaptive comparative judgement (ACJ) in the cloud-based platform. ACJ is an assessment methodology where the judgers compare two pieces of student work and identify which one of them is better, without saying how much better it is. Their decision is based on quality of the work. This judgement process is repeated until a rank order of student work has reached reliability greater than 90% and is professional consensus among judgers (Pollitt 2011 2012). Professors

Richard Kimbell and Professor Kay Stables, whose work has had a great influence on mine, introduced ACJ to technology education. ACJ was selected as the most appropriate approach to investigate what criteria for success teachers emphasized during assessment.

The pair engine software for ACJ generates quantitative data. However, to identify the motives behind teachers' choices, teachers were asked to describe the reasons for their choices. These quantitative think-aloud protocols were recorded and transcribed into protocols. They provided valuable insights on the rationale for each choice.

What I Found Out

The results confirmed that teachers use complex assessment practices in different ways to provide learning opportunities for student progress.

The study examined the design and utility of the IDP in primary technology (Hartell 2014). We found that the IDP is unsuitable for use as a formative assessment in technology education because it fails to provide information regarding students' current or future progress in the subject. There are two reasons for this failure. First, the long time span between gathering evidence of the learning and putting it into practice (Hirsh 2011) impedes theoretical use of the IDP. Second, it lacks information regarding technology, as technology is often missing in the schools' standardized IDP templates. This omission severely impedes teachers' ability to use the IDP realistically.

This particular study also provided interesting results regarding student achievement in technology education prior to Year 9 (Year 9 grades were the only statistics available at a national level when this study was conducted). The study revealed a remarkably high level of achievement among primary students, with only a few students failing to achieve the expected outcome. This contradicts the prevalent view of lack of technology instruction, especially in primary schools in Sweden. However, due to changes in regulations, teachers nowadays must complete the IDP once a year for Years 1 to 5, i.e. prior grading instead of twice a year all through compulsory school. The National Agency for Education has also revised their instruction governing the document and provided a support template for the IDP, which includes technology.

Policy and follow-up documents are important; however, for students, what happens in the classroom is far more important. Hence, the second study sought to rectify the gap in knowledge regarding teachers' classroom assessment practices. This study is documented in the paper, *Looking for a glimpse in the eye* (Hartell 2013), which was so titled because one of the teachers, Karl, explained that he could tell whether his pupils had understood *from that glimpse in the eye*.

Analysis of the data collected in classrooms is evidence that these teachers used assessment for learning and for encouraging pupils' progress. The observers noted that the teachers used all five strategies for formative assessment (Black and Wiliam

1998). The observations confirmed previous results (e.g. Black 2008; Kimbell 2007; Moreland et al. 2008) and that these strategies enable teachers to infer evidence about their pupils' progress (Black et al. 2004; Black and Wiliam 2009), including probing questions, listening, interpreting responses, waiting for students to clarify their thoughts, looking for body language cues and encouraging pupils to think. It was clear that these two teachers did everything in their power to enhance student learning. However the direction of where they were going was not explicit. Both teachers demonstrated belief in their pupils' abilities to learn and placed great emphasis on the language of technology and on Swedish language in general. Not so surprising as the majority of their students had Swedish as their second (or third) language.

The results showed limited affordances for teachers to make every pupil prosper. Affordance here refers to the quality of the environment, including both material and nonmaterial prerequisites, which enables teachers to perform their jobs. It is concerning that results show clearly that these teachers did not have a support system and an education that assisted them. They depended entirely on their own prior experience as teachers and so did their pupils. It is perhaps not unusual for pupils to be dependent on their teacher; however to be solely dependent on one solitaire teacher, this must be questioned in terms of equity. Equity in education is highly emphasized in the Swedish Educational Act. For example, neither teacher had received training in technology nor did they have access to a peer support network, which is largely needed according to research. In addition, they faced severe organizational deficiencies such as lack of tools and equipment; not only did they bring teaching equipment from home but they also set it up, which was an inefficient use of their time. There was also a lack of teacher education, documenting practice, in-service training, professional development, documenting templates, teaching materials and teaching resources, which was supported by prior research (Harrison 2009; Pettersson 2009) as highly relevant for validity and reliability of assessment. Thus, technology teachers in Sweden are currently not well supported by the environment.

As mentioned earlier, these teachers lacked formal training in technology education, as do the majority of Swedish teachers. This is of concern because according to the new legislation introduced in 2011, only trained teachers can secure permanent teaching positions and award grades to students. While we do not yet know how teacher training affects student achievement, it is clear that the one indicator that does affect student achievement is teachers' self-efficacy (Hattie 2009). The study selected the perspective of self-efficacy because it influences student achievement and is highly context bound. Therefore the third study narrowed the scope down and focused on teachers' self-efficacy in assessment in technology.

The results from this third study indicated that training in technology affected teachers' self-efficacy for assessment in technology in a positive way. We found significant differences between teachers who had received technology training compared to those who had not, in terms of their self-efficacy regarding informing students of their expectations and in describing students' level of knowledge in writing and reporting using mandatory curriculum documents.

However, the fact remains that most technology teachers in Sweden lack subject-specific training and often any form of teacher training; the consequences of this are unknown. The certification process, which includes that teachers have to have a teacher degree from university to be able to get a certificate, and the fact that only teachers who have a certificate are allowed to grade students, which is now taking place in Sweden, may be a step in the right direction, but it is not sufficient. For example, for grading Swedish students, teachers do not need a certificate in technology education to grade the students; any teacher certificate will do. And, the subject-specific content in technology, which includes a wide range of topics, also changes quickly, and perhaps more importantly, there is a lack of knowledge regarding what constitutes learning in technology. Lifelong learning in subject-specific training, which focuses on both content and pedagogy, must be embedded into teachers' assessment practices. Although both groups indicated lower self-efficacy in terms of assessment for grading, teachers without a teacher-training degree but with a technology background showed the highest self-efficacy. This group also did not use the national and local curriculum as the basis for their teaching practice, unlike their peers who had received teacher training. This finding raises questions in terms of curriculum alignment, especially because there is a large variance in awarded grades.

In classrooms, teachers make inferences based on nods, winks and glimpses. This thesis has contributed to the field by showing what teachers focus on and the criteria for success that primary teachers emphasize during assessment. Teachers agreed upon the importance of students providing evidence of learning so they can see the narrative of student progress. They also agreed that students must complete their tasks. One interpretation of this could be that completed tasks provide more evidence for learning. Another interpretation could be that completing tasks is a virtue. Regardless, it puts high demands on the design of the task that students must perform and the circumstances in which they work, which highlights the process of planning instruction. Pupils must have opportunities to learn in an environment where teachers encourage risk-taking and learning from mistakes as well as sufficient time to finish their tasks. These learning opportunities must be included in instruction. Pupils must not just be left alone to figure things out on their own, as has also been reported by the Swedish School Inspectorate (Skolinspektionen 2014).

This broad thesis covered many aspects of teachers' assessment practices. The four sub-studies in combination with literature concluded that affordance for teachers' assessment practices must increase in order to bridge teaching and learning in technology education.

How This Might Be Used to Improve Teaching and Learning

This thesis demonstrated that teachers use assessment in different ways and that the educational environment in Sweden is not adequately supportive of teachers' efforts to ensure the progress of their pupils. Formative assessment is necessary for bridging teaching and learning in technology; thus, affordance for teachers' assessment practices must increase. The complexity of this is acknowledged, and I suggest some measures below.

Formative Assessment and Affordance to Bridge Teaching and Learning

Short-cycle formative assessments undertaken minute-by-minute in the classroom setting have great potential to provide opportunities for student learning when subsequently incorporated into everyday classroom instruction practices (Wiliam 2009). However, there are a lot of misconceptions regarding formative assessment (Bennett 2011; Hirsh and Lindberg 2015).

Formative assessment is based on two fundamentals: (1) pupils don't learn everything they are taught, and (2) good teaching starts from where the students are. Formative assessment is about eliciting, interpreting and inferring from the evidence of learning and using this information to meet the needs of pupils. Teachers must ensure their pupils' progress according to the regulations that govern compulsory schooling. Teachers must also invite pupils to become owners of their own learning in technology. This is highly subject-specific and context bound, puts a lot of pressure on teachers and demands both assessment-literate and self-efficacious teachers who are supported by the environment in which they are teaching. My research has found that teachers are pretty much left on their own to cover up the deficiencies in the system level—an inefficient loneliness.

Increasing Affordance

Increasing affordance, i.e. the quality of the environment in which teachers are situated, for teachers' assessment practices, for example, by focusing on teaching and learning, can diminish the evident lack of support for assessment practices. Teachers can utilize their effort and time wisely and efficiently if they can support each other in planning and designing the procedures to support and draw inferences from pupils' learning to inform instruction.

From an assessment point of view, it is important to embed into the teaching process the five key strategies for formative assessment: clarifying and sharing learning intentions and criteria for success; engineering effective classroom discussions, questions and learning tasks; providing feedback that progresses learners; activating students as owners of their own learning; and activating students as instructional resources for one another (Leahy et al. 2005). One way to do that is to follow Winnie the Pooh's advice: 'It is better to know what to look for before you start looking for it'. Learning is complex and it is quite clever to have some ideas on where you are going. What are your students supposed to learn? Do you know what success looks like? Do you know how to share that success in a pupil friendly way? Pupils need to know where they are going, and they need to know when they have arrived (Black and Wiliam 2009). All pupils benefit from teachers being clear and transparent in their communication regarding goals and criteria for success; however, the low achievers benefit the most (Jönsson 2010; Moreland et al. 2008). Clearness and transparency in communication place demands on teachers' understanding of the subject (Moreland et al. 2008). As with other school subjects (Jönsson 2010), alignment and transparency in assessment are important in the technology classroom too.

Another concern involves the environment of the technology classroom, which needs to be safe enough for students to reveal their concerns and ideas (Black 2008; Moreland et al. 2008; Wiliam 2016). Within the Swedish context, these factors are problematic. Even though assessment was not the focus of the review undertaken by the SSI (Skolinspektionen 2014), they still put forth concerns about how students' knowledge is assessed and graded with regard to technology. So what about the context where your students are situated? Do you know how to plan ahead what activities need to be included to meet these success criteria? This research indicated the importance of including questions to pose and check pupils' progress during the learning journey, instead of focusing assessment afterwards by providing suggestions on what the student should have done or not, to make even greater progress. Therefore, include questions in your lessons plans, show them worked exemplars on what success can look like and give time and space for your students to ask questions. Also the importance of feedback is what receiver do with feedback, so include time and space for learners to work with the feedback that you provide to them and also what they provide to you—in terms of them asking questions or answering question and other feedback activities. And share these with your colleagues.

It is also necessary for teachers to improve their self-efficacy and assessment literacy, both in terms of pedagogy and subject-specific knowledge. Rohaan et al. (2012) investigated the possible relationships between subject-content knowledge, pedagogical-content knowledge (PCK) and the attitudes and self-efficacy of Dutch primary teachers. They demonstrated that developing subject-content knowledge and PCK among teachers improves their instructional self-efficacy and positively affects their attitudes towards technology education. Correspondingly, increased instructional self-efficacy and positive attitudes increased the frequency of technology education activities. The quality of those activities included is also important. However, this creates a purposive cycle for primary technology teachers, where their experiences of teaching technology strengthen their PCK and subject-content knowledge, leading to increased self-efficacy, improving their attitudes towards the subject and enhancing the quality of primary technology education.

A perspective of lifelong learning must be embedded into teachers' assessment practices, and it must involve subject-specific training and focus on both content and pedagogy. Subject-specific training has a positive effect on teachers' self-efficacy of assessment. However, self-efficacy is highly context bound and may be related to teaching a particular content/topic or a particular year or group (Hartell 2017). The subject-specific content in technology changes quickly, and there is a lack of knowledge regarding what constitutes learning in technology. In addition, support to interpret curricula and perhaps even suggest ways to instruct is required. For example, the Swedish National Agency for Education has produced support material for teaching and assessment in technology (Skolverket 2011), which may be used as navigational tools for education. Though promising, the material must be put into practice to be useful, and teachers need affordances to do so (Palmer et al. 2015; Nilsson 2008).

From investigating student-teacher progress during teacher education, Nilsson (2008) emphasized the importance of experience and practice in how to teach in order to develop self-efficacy in specific content areas. Rohaan et al. (2012) also suggest that teachers themselves should experience hands-on activities to increase confidence and self-efficacy by gaining relevant subject-content knowledge and using it to engineer learning opportunities for students, including pedagogical approaches relevant to technology education. Teachers should also continue to learn about the nature of teaching technology, including designing provoking questions and repertoires of explanations as well as recognizing common misconceptions.

Working Together at the School Level

At the school level, teachers must discuss and plan lesson activities that focus on what students will be learning and explain their expectations clearly to the students. Teachers must provide time for students to learn from mistakes. Teachers can also form support groups with their colleagues to share equipment, lesson plans, etc. and save time. Please forgive me for being a bit preachy here, but please do not spend time documenting if the information is not used in the classroom to meet learners' needs. Question authority when they oblige you to. However, do ensure that you document evidence of learning to measure students' progress and use it.

In designing teaching and learning opportunities for their students, teachers need to start from where the learners are and provide the necessary affordances to move them forward on their learning journey. For this to happen, teachers need to know when to push and when to hold back, using evidence of learning to adapt what happens in the classroom to best meet their students' needs in each individual context.

However, teachers cannot do this alone. The quality of leadership is significant in influencing efficacy by organizing the milieu in which teachers work and thereby contributing to teachers' self- and collective instructional efficacy (Goddard et al. 2000; Goddard et al. 2015; Hattie 2012; Pettersson 2009; Timperley 2011). Therefore, it is very important to invite school leaders from every level of the educational system to develop strategies to provide affordances for teachers' assessment practices, so they can embed assessment in their practice to bridge teaching and learning.

Working Together on a Global Scale

The context of the technology subject involves different meanings across the world. It is particularly interesting to investigate assessment in different contexts and with different purposes in Sweden and elsewhere. What are the important factors of technology education? How do these factors relate to other subjects? Technology is often combined with science, engineering and mathematics in the growing science, technology, engineering and mathematics (STEM) movement. What is subject-specific and what is general within the realm of STEM? What is specific and what can be synergetic to increase student capabilities and interest within these areas, in all forms of education? This is an international need, and international collaboration is necessary. International collaboration can enable teachers to be attuned with what, how and when technology education can benefit or support student learning. For example, the use of digital tools such as ACJ can clarify learning intentions and the criteria for success for future students and build assessment literate and self-efficacious teachers. Build upon the work of Sweller et al. and produce worked exemplars that show learners what quality can look like and perhaps be used to grow a nose for quality and share these exemplars across the globe.

What Else Should Teachers Know

Teachers should be well informed in terms of formative assessment and how it can be embedded into teaching practice. Teachers can begin by reading *Embedded Formative Assessment* by Dylan Wiliam and Siobhan Leahy (Leahy and Wiliam 2010), and perhaps an even more efficient way to embed formative assessment in everyday learning activities is working with peers. Implementing these recommendations requires that teachers are able to plan instruction, develop forms of collaboration and share knowledge while taking responsibility for bridging teaching and learning.

Much research and many teachers have documented experience of the implementation of formative assessment, and it is important to build on this experience (Hartell 2015; Hartell and Ahlkvist 2014; Hartell et al. 2015). Assistance with formative assessment has been well received by teachers as it is focusing on a key activity of schooling—classroom instruction. School principals and local school authorities must design an environment that supports teacher learning and provide affordances to release the power of assessment to bridge teaching and learning for the sake of pupils.

Meaning of Assessment

'If you are serious about raising student achievement then you have to be focusing on AfL [Assessment for Learning], and if you are not focusing on AfL you are probably not serious about raising student achievement' (Wiliam 2009, p. 34). This quote summarizes research and highlights the important role of AfL/formative assessment.

Formative assessment is an integral part of bridging teaching and learning, ensuring pupils' growth. However, the importance of affordance for teachers' assessment practices cannot be emphasized enough. Teachers need to be in an environment that supports them, where they can grow their assessment literacy and self-efficacy. Through a play with words, an important understanding of assessment is put forward. The word assessment originates from the Latin assidere, meaning to sit by. Playing with the term 'pupil' by translating it back and forth English to Swedish, interesting things happen. First, you get student or elev (where elev originates from the French word élever meaning to raise, uplift or exalt). Also, pupil can also be translated to mean a part of the eye, in other words, the medical term 'pupil'. This may refer to the apple of one's eye or 'ögonsten' in Swedish, or even a sweetheart. So to me working with assessment is as if I were sitting beside someone very dear to me, and having the ambition to elicit evidence of learning, put these evidence of learning into action in the classroom to better meet the pupils' or rather my sweathearts' needs.

References

Bennett, R. E. (2011). Formative assessment: A critical review. *Assessment in Education: Principles, Policy & Practice, 18*(1), 5–25. https://doi.org/10.1080/0969594X.2010.513678.

Black, P. (2008). Formative assessment in the learning and teaching of design and technology education: Methods and techniques. *Design and Technology Education: An International Journal, 13*(3), 19–26.

Black, P., & Wiliam, D. (1998). Assessment and classroom learning. Assessment in education: Principles. *Policy & Practice, 5*(1), 7–74. https://doi.org/10.1080/0969595980050102.

Black, P., & Wiliam, D. (2009). Developing the theory of formative assessment. *Educational Assessment, Evaluation and Accountability, 21*(1), 5–31. https://doi.org/10.1007/s11092-008-9068-5.

Black, P., Harrison, C., Lee, C., Marshall, B., & Wiliam, D. (2004). Working inside the black box: Assessment for learning in the classroom. *Phi Delta Kappan, 86*(1), 8–21. Retrieved from https://www.researchgate.net/publication/44835745_Working_Inside_the_Black_Box_Assessment_for_Learning_in_the_Classroom.

Goddard, R. D., Hoy Wayne, K., & Woolfolk Hoy, A. (2000). Collective teacher efficacy: Its meaning, measure, and impact on student achievement. *American Educational Research Journal, 37*(2), 479–519.

Goddard, R., Goddard, Y., Sook Kim, E., & Miller, R. (2015). A theoretical and empirical analysis of the roles of instructional leadership, teacher collaboration, and collective efficacy beliefs in support of student learning. *American Journal of Education, 121*(4), 501–530. https://doi.org/10.1086/681925.

Harrison, C. (2009). Assessment for Learning. A Formative Approach to Classroom Practice). In A. Jones & M. deVries (Eds.), *International handbook of research and development in technology education* (pp. 449–459). Rotterdam: Sense Publishers.

Hartell, E. (2013). Looking for a glimpse in the eye: A descriptive study of teachers' work with assessment in technology education. In I.-B. Skogh & M. J. De Vries (Eds.), *Technology teachers as researchers: Philosophical and empirical technology education studies in the Swedish TUFF research school* (1st ed., pp. 255–283). Rotterdam: Sense Publishers.

Hartell, E. (2014). Exploring the (un-) usefulness of mandatory assessment documents in primary technology. *International Journal of Technology and Design Education, 24*(2), 141–161. https://doi.org/10.1007/s10798-013-9250-z.

Hartell, E. (2015). The cared for teacher. In J. Evers & R. Kneyber (Eds.), *Flip the system: Changing education from the ground up* (1st ed., pp. 241–246). London/New York: Routledge.

Hartell, E. (2017). Teachers' self-efficacy in assessment in technology education. In M. J. de Vries (Ed.), *Springer international handbook of education: Handbook of technology education* (1st ed., pp. 1–16). Cham: Springer. https://doi.org/10.1007/978-3-319-38889-2_56-1.

Hartell, E., & Ahlkvist, J. (2014). En kortfattade beskrivning om hur en skola i Haninge kommun har arbetat med att implementera formativ bedömning hållbart i verksamheten. In *Bedömning för lärande, BFL, för praktik och forskning* (pp. 1–3). Stockholm: IFOUS Retrieved from http://www.ifous.se/bfl-konferens/.

Hartell, E., Holmberg, S., & Åkesson, J. (2015). Att leda framåt och inte till leda. In *Att leda för BLF; forskning och praktik möter varandra* (pp. 1–2). Stockholm: IFOUS Retrieved from http://www.ifous.se/bfl-2015/.

Hattie, J. (2009). *Visible learning: A synthesis of over 800 meta-analyses relating to achievement*. London: Routledge.

Hattie, J. (2012). *Visible learning for teachers Maximizing impact on student learning*. Exeter: Routledge.

Hirsh, Å. (2011). A tool for learning? An analysis of targets and strategies in Swedish individual educational plans. *Nordic Studies in Education, 31*(1), 14–30.

Hirsh, Å., & Lindberg, V. (2015). *Formativ bedömning på 2000-talet- en översikt av svensk och internationell forskning*. Sweden: Stockholm. Retrieved from https://publikationer.vr.se/produkt/delrapport-fran-skolforsk-projektet-formativ-bedomning-pa-2000-talet-en-oversikt-av-svensk-och-internationell-forskning/

Jönsson, A. (2010). *Lärande bedömning*. Malmö: Gleerups.

Kimbell, R. (2007). Assessment. In M. de Vries, R. Custer, J. Dakers, & G. Martin (Eds.), *Analyzing best practices in technology education* (pp. 247–258). Rotterdam: Sense Publishers.

Leahy, S., Lyon, C., Thompson, M., & Wiliam, D. (2005). Siobhan Leahy, Christine Lyon, Marnie Thompson, and Dylan Wiliam. Quality Assurance, 63(November), 1–7. Retrieved from http://www.ascd.org/publications/educational-leadership/nov05/vol63/num03/Classroom-Assessment@-Minuteby- Minute,-Day-by-Day.aspx

Leahy, S., & Wiliam, D. (2010). *Embedding formative assessment-a professional development pack for schools*. London: Specialist Schools and Academies Trust.

Lindström, L. (2006). Creativity: What is it? Can you assess it? Can it be taught? *The International Journal of Art & Design Education, 25*(1), 53–66.

Moreland, J., Jones, A., & Barlex, D. (2008). *Design and technology inside the black box assessment for learning in the design and technology classroom*. London: GL Assessment.

NAE. (2009). *Allmänna råd och kommentarer. Den individuella utvecklingsplanen med skriftliga omdömen*. Skolverkets allmänna råd 2008. Revised 2009. Statens Skolverk (NAE).

Nilsson, P. (2008). *Learning to Teach and Teaching to Learn Primary science student teachers' complex journey from learners to teachers*. Norrköping: Linköping university.

Palmer, D., Dixon, J., & Archer, J. (2015). Changes in science teaching self-efficacy among primary teacher education students. *Australian Journal of Teacher Education*, (12), 40. https://doi.org/10.14221/ajte.2015v40n12.3.

Pettersson, A. (2009). Bedömning- varför, vad och varthän? In L. Lindström & V. Lindberg (Eds.), *Pedagogisk bedömning* (2nd ed., pp. 31–42). Stockholm: Stockholm universitets förlag.

Pollitt, A. (2011). Comparative judgement for assessment. *International Journal of Technology and Design Education, 22*(2), 157–170. https://doi.org/10.1007/s10798-011-9189-x.

Pollitt, A. (2012). The method of adaptive comparative judgement. *Assessment in Education: Principles, Policy & Practice, 19*(3), 281–300. https://doi.org/10.1080/0969594X.2012.665354.

Rohaan, E. J., Taconis, R., & Jochems, W. M. G. (2012). Analysing teacher knowledge for technology education in primary schools. *International Journal of Technology and Design Education, 22*(3), 271–280. https://doi.org/10.1007/s10798-010-9147-z.

Skolinspektionen. (2014). *Teknik – gör det osynliga synligt Om kvaliteten i grundskolans teknikundervisning*. Stockholm: Skolverket.

Skollagen Svensk författningssamling 2010:800. (2013). Retrieved from http://www.riksdagen.se/sv/Dokument-Lagar/Lagar/Svenskforfattningssamling/Skollag-2010800_sfs-2010-800/?bet=2010:800

Skolverket. (2011). *Diskussionunderlag för grundskolan Diskutera kursplanen i ämnet teknik*. Retrieved from http://www.skolverket.se/publikationer?id=2547

Timperley, H. (2011). *Realizing the power of professional learning*. Chippenham: Open University Press.

Wiliam, D. (2009). Assessment for learning: Why, what and how? In *An inaugural professorial lecture by Dylan Wiliam*. London: Institute of Education University of London.

Wiliam, D. (2016). *Leadership for teacher learning. Creating a culture where all teachers improve so that all students succeed* (1st ed.). West Palm Beach: Learning Sciences International.

Chapter 9
Approaches to Planning that Encourage Creativity

Mary Southall

Abstract Planning is a crucial part of the teaching and learning process yet is often a stage that is either overlooked or deliberately neglected. The research on which this chapter is based investigates the approaches used by technology teachers to plan how teaching, learning and assessment activities work together to ensure students have the opportunities to demonstrate intended learning. Particular focus is placed upon creativity and the development of creative skills in relation to the planning approach.

The research adopted a mixed methodology approach to analyse planned and actual teaching and learning activities and indicated a strong causal relationship between the planning approach used and the creative output. The dominant approach to planning produced learning outcomes that are both easy to measure and easier to teach but did not support teaching and learning of cognitive skills, such as reflection and creativity. In the light of these findings, this chapter considers how reflection and creativity might be supported in the classroom.

Key Questions That Need To Be Asked About Teaching and Learnng and Why They Are Important?>

Learning is a personal process that can occur through a range of planned learning activities, such as collaborative groupings, peer discussions, self-reflections and focused tasks. The teacher needs to plan learning activities that provide opportunities for students to learn the intended knowledge or skills and develop an understanding of the learning focus. In order for both teacher and learner to assess the intended learning and allow for necessary modifications to be made, feedback opportunities, through planned assessment activities which focus on the progress of

M. Southall (✉)
Western Sydney University, Sydney, Australia
e-mail: mary@attainmentpartnership.org.uk

P J. Williams, D. Barlex (eds.), *Explorations in Technology Education Research*, Contemporary Issues in Technology Education,
https://doi.org/10.1007/978-981-13-3010-0_9

the learner and the effectiveness of the planned learning activity, need to be embedded into the learning activity.

The complexity of the learning process means it is something that cannot always be easily isolated or exemplified. Learning progress or the process of learning, by its very nature, is difficult to 'confine' to a single location or moment in time (Hewitt 2008) as it involves psychological, biological and social conditions and many processes and dimensions, learning types, learning barriers and specific internal and external conditions (Illeris 2009).

In order to successfully plan for learning progress, the teacher needs to plan for what has to be learned; how it will be learned, that is, the learning activity; and the form it will take in terms of the learning outcome. It is then necessary to assess the learning against predetermined criteria. The planning process is crucial in the process of teaching and learning, and research into this area tends to be predominantly focused upon how novice teachers learn to plan and master the process of transforming and adapting the curriculum into learning experiences and opportunities (Clark and Lampert 1986). Research into the planning processes and procedures used by more experienced teachers is often disregarded, particularly in relation to subject specialisms.

The process(es) of planning for learning was the main focus of my research which concentrated upon two key questions:

- How do technology teachers plan to achieve the intended learning outcome(s)?
- What methods are used to capture and gather evidence of students' learning in technology?

My main interest lays in two key stages of the teaching-learning process. Firstly, how teachers plan for the development of creative skills in their students, focusing on the methods teachers used to translate the specific learning into a learning activity that would allow a student the opportunity to demonstrate the specific learning, how teachers evidenced this and whether the evidence could be suitably assessed. And secondly, the demonstration of learning or the learning outcome and the form the teacher planned for the learning outcome to take. Clearly, the form relates to the nature of the learning activity, for example, a written text, a 3D realisation, a videoed role play or a PPT presentation.

Generally, it is the teacher (and not the learner) who identifies and then dictates the form of the learning outcome during the planning process. The teacher then assesses the learning outcome that provides the evidence of learning or demonstration of learning by the student, with the assessment criteria assisting in the determination of the evidence of learning.

Within such a teacher-led, teacher-dominated process, where does the development of creativity need to be emphasised? This research aimed to provide greater understanding of the process(es) used by technology teachers to plan for technology skills, knowledge and understanding and to highlight effective planning approaches in terms of creative development.

How I Addressed My Research Questions

When investigating such a dynamic and complex process as the teaching-learning process, it is imperative that the research is well planned, thorough and justifiable. The research study was divided into three distinct yet interrelated studies: study 1 focused upon the planning processes that are the 'pre active stage'; study 2 involved investigating the teaching-learning activities and was termed the 'inter active stage'. This study focused upon the learning outcomes produced during the lesson and formative assessment activities during the lesson. Study 3 concentrated upon the learning outcomes, the 'post active' stage. Figure 9.1 presents a graphical representation of the three studies.

Mixed methods research involves collecting and analysing quantitative and qualitative data. Analysing both types of data allows the data, and the consequent findings, to intertwine and to 'talk to each other' throughout the research study (Datta 2001: 34), an approach termed crossover tracks analysis by Greene (2008).

Study 1 investigated the planning processes and procedures used by technology teachers when planning lessons. It addressed the question, how do technology teachers plan to achieve the intended learning outcome(s), and aimed to provide insights into the processes teachers employ when designing learning experiences in technology.

Study 1, part 1, involved collecting and analysing technology teachers' written descriptions of their proposed teaching-learning processes in the form of lesson plans. Written before the lesson, these teaching-learning plans offered a description of the predicted event, providing evidence of the intended learning statements, how the teacher proposed to structure the learning activities or episodes to achieve the intended learning and by what means the teacher intended to reveal and capture the learning through the learning outcomes.

The lesson plans were analysed using a thematic analysis approach primarily to identify any broad patterns or trends in the data (see Guest (2012) for extensive discussion on this method). Thematic analysis is not tied to any particular epistemology or discipline and is a process that can be used with many kinds of qualitative

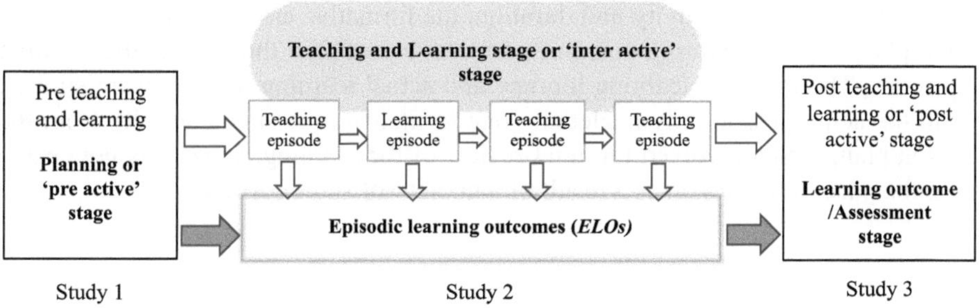

Fig. 9.1 The teaching-learning process model used in this research study

data. A degree of flexibility was necessary at this stage of study, as the lesson plans were different in their form, formatting, overall design and emphasis.

Seven schools participated in Study 1, part 1, and lesson plans were gathered from years 7, 8 and 9 technology lessons. A total of 40 lesson plans were sent to the researcher within the requested period, a response rate of 57%.

Study 1, part 2, involved emailing each of the participating teachers and requesting them to answer the question, 'How do you plan for your lessons?'. The email simply asked for a response, via email, to the question within a week of receiving the request. A total of 23 responses were collated and analysed using a thematic approach.

Study 2 provided a direct comparison between the lesson intended to be delivered by the teacher and the actual lesson delivered by the teacher. It examined how the learning outcome was produced and how it was captured. With a particular emphasis on formative assessment activities planned and used by the teacher to reveal the learning, Study 2 aimed to answer the question, 'what methods are used to capture and gather evidence of students' learning in technology?'

Study 2 used a series of semi-structured classroom observations as its main research tool. During each observation, the aim was to identify the students' possible learning throughout the lesson. It was considered a relatively low inference observation, that is, the lesson observation focused upon observable facts and events and involved a low degree of subjectivity, such that the researcher was required to 'notice' but not to 'judge'.

Through observations, learning can be examined in relation to both the intended learning statements and the students' learning during the actual learning episode. When the observations are extended to enough cases, generalisations could be made, and it was hoped this would be the case in Study 2.

Study 3 compared the intended learning episodes with the actual learning episodes, focusing upon learning outcomes and formative assessment, and was designed to provide data on the usefulness of the planning process employed, in relation to the progression of students' learning.

A graphical representation (see Fig. 9.2) allowed for a direct visual comparison of pre- and post-learning opportunities.

The upper graphic represents the actual lesson with the 'blocks' representing the focus of the learning activity and duration, the formative assessment activities that took place and the learning outcomes that were produced throughout the learning episodes. The intended learning journey and actual learning journey were similar, with the actual lesson only neglecting one formative assessment activity from the lesson plan, which involved the final review of the learning. What was noticeable was the pace of the lesson as a result of a number of short episodes.

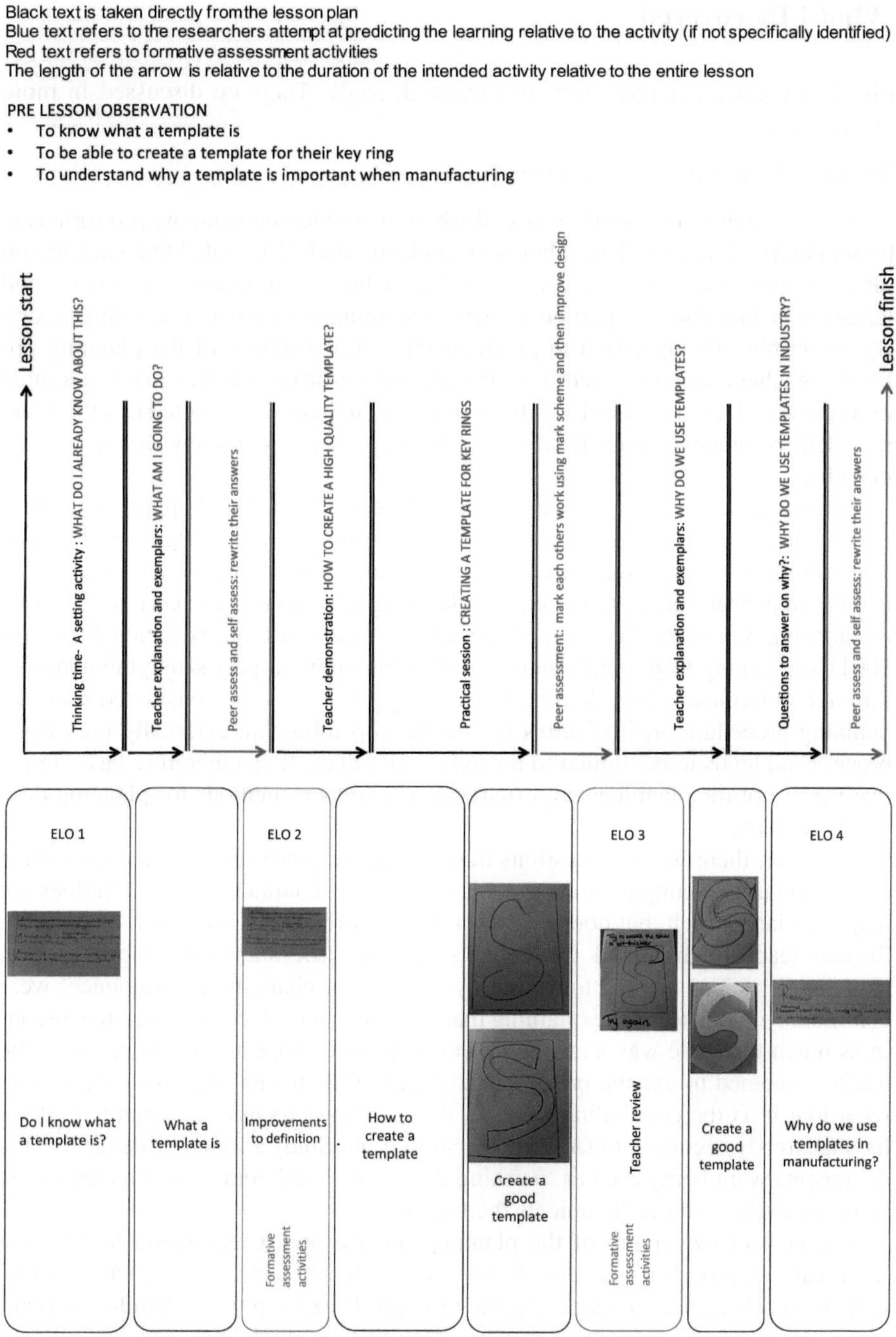

Fig. 9.2 Graphical representation of intended and actual classroom-based learning

What I Discovered

Five key themes emerged from this research study. They are discussed in more detail below:

Theme 1: Formal planning procedures

This research study found the use of whole school lesson planning pro formas to be standardised in secondary schools in England. Study 1 revealed that such lesson planning proformas were generally designed by senior teachers/managers and tended to be based upon a particular view of learning or a particular learning theory. By prescribing the approach to planning through the design of the planning pro forma, teachers were removed from the planning process, which became a simple procedural task (Milkova 2014). The findings from Study 1 supported this view and included comments from teachers such as, 'There is no ownership of the planning'.

The whole school planning pro formas followed a 'dominant' approach to planning for learning (Tyler 1949) in terms of their overall design and structural layout. The 'dominant' approach is based upon the formulation of intended learning statements and a clear notion of the required learning outcome(s). The current literature on planning processes deems the 'dominant approach' or 'objective model' (Elliott 2001) to planning as primarily managerial in intention, largely satisfying administration requirements. The 'dominant' planning process can be considered a formal planning procedure, as it requires the teachers to follow an externally prescribed process and tends to be limited to formal observations. It can therefore be assumed that the use of the planning pro forma does not support methods for planning day-to-day lessons.

However, there were suggestions that the planning pro forma was useful when considered as a 'quality assurance' tool, with teacher comments such as 'It does not help me plan as such, but does help as it ensures everything is ok with the lesson'. Several teachers discussed the benefits of using the standardised pro formas. Comments such as helping to 'order my thoughts or changing the sequence' were common, with one teacher explaining that the proforma helped her 'craft the lesson' in as much the same way as using PowerPoint slides appeared to. In practice, the teachers seemed to use the pro forma to 'finalise' their thinking about the lesson, regarding it as the concluding stage in the planning process, the cognitive phase having already occurred. In fact, neither Study 1 nor Study 2 found evidence that the documents were being used as 'working documents' and notations or amendments were not evident on the documents themselves.

The lack of ownership of the planning process and a requisition to follow a 'dominant' approach to planning have clearly affected technology teachers' planning approaches and, consequently, the teaching-learning process. Whilst the planning process required teachers to translate syllabus guidelines and institutional expectations, teachers also incorporated their own beliefs and ideologies of education and the subject specialism into plans for teaching lessons. It was evident throughout the research study that teachers' own beliefs about technology teaching and learning were combined with their professional knowledge when planning learning activities (Shavelson and Stern 1981; Calderhead 1984), albeit within a prescriptive framework.

Theme 2: Informal planning processes

Findings from Study 1 and Study 2 revealed that, in practice, teachers used a variety of informal approaches to planning. Teachers tended to plan at various times of the day by mentally focusing on issues that need their attention (see Hagger and McIntyre 2006). These research study's findings support this claim, with a myriad of informal approaches being highlighted by the technology teachers, including group discussions, talking to wife/technician/colleague, scribbling notes, mind mapping and thinking in the car. Such informal planning approaches appear to be very personal, with several teachers having devised bespoke methods to suit them and their particular circumstances.

It was interesting to note that teachers found it difficult to articulate their planning process in terms of the mechanisms or detailed processes involved. Planning tended to be stimulated by a 'thought' often triggered by an inspiration, for example, a product or television programme, which was then translated into a learning activity or learning journey. It was a clear and recurring theme in teacher responses that there was a need for a creative aspect in the planning process.

The use of PowerPoint presentation slides to formulate the learning journey was a common method of informal planning, often justified by the degree of flexibility and creativity PowerPoint slides provided. The software allowed teachers to develop and clarify their thinking about the learning journey, translating vague ideas into an effective learning journey. Ideas are formed and, after elaboration, developed into mental plans or images and act as classroom scripts or guides (Morine-Dershimer and Vallance 1976). Several teachers used the common analogy of 'a journey' (Peters 1965) when discussing their planning processes. For example, one teacher stated, 'I start with the aim of the lesson and what I need them to have learnt by the end and then think about how I can get there'. PowerPoint slides provided the teachers with a clear 'image' of the learning journey, slide by slide, and a mechanism for teachers to visualise the intended lesson and learning journey. In this regard, PowerPoint slides also allowed the teacher to rearrange the learning journey episodes in order to ensure a logical, progressive learning journey.

The use of PowerPoint slides when planning for learning exemplifies some of the current research on planning relating to 'planning and images' borne out in the literature review (Clandinin 1985, 1986). Such an approach seems to be characterised as a mental rehearsal of ideas and knowledge about the students, the school and the curriculum. Without the notion of planning in the form of visualisation, it is difficult to anticipate the ways in which what has been planned may unfold in the classroom. More experienced teachers approached planning as the anticipation of what might happen rather than their determination of what would happen, that is, planning as visualisation rather than planning as a template (Mutton et al. 2011). 'Novice planning processes' identify this 'image making' as a significant feature and a sign of growing confidence and belief in teaching and planning structures (John 2006).

Theme 3: The 'learning focus'

The majority of teachers discussed ways of deciding upon a 'learning focus' at the start of the informal planning process. This is despite the fact that 'learning' was not the key stimulus or motivation for the planning process and was not the first response to the survey question 'how do you plan for learning?'. Teachers specified minimal details in regard to processes for formulating learning, offering far more information on planning for the resources to support the learning intention. As suggested by Tsui (2002), teachers consider materials and resources, students' interests and abilities as the first aspects in the planning process.

Theme 4: The learning journey

All three studies revealed a lack of detail in relation to the teaching, learning and assessment activities evident in the planned episodes/activities. This appeared to encourage discreet activities that could be isolated in terms of learning. Teaching strategies were either not specified or were vague in their description, and learning was often defined through what the students were going to do during the lesson. Details on the formative assessment activities were minimal, often stating 'peer assessment' or 'mini white boards', suggesting that the teachers had specific experience in these lesson activities. In relation to the planned question and answer sessions, there was no information on what questions were to be asked, suggesting that teachers 'improvised' to address the perceived needs of the class (McAlpine et al. 1999).

Through the use of whole-school lesson proformas, formative assessment activities were integrated into the learning journey, and as a result, there was clear evidence that teachers were planning formative assessment activities alongside the teaching and learning activities and not as a 'bolt on' at the end of the plan (Sondergeld 2010: 77). This is a benefit of whole-school planning proformas. Surprisingly, the survey results in relation to the informal planning processes provided no evidence of formative assessment being included or incorporated into informal planning, and there was also no mention of the need to review the learning or ascertain the level of learning.

The value of formative assessment, for both the teacher and the learner, has been widely discussed in the current formative assessment literature. Formative assessment activities were generally well planned into the learning journeys, six of the lessons involved at least one formative assessment episode. The majority of the formative assessment strategies involved 'peer' or 'self'-assessment activities. It became apparent whilst observing the lessons that the number of planned formative assessment activities significantly reduced in the 'interactive' phase. Twelve formative episodes were planned into seven lesson plans, and this was reduced to eight episodes during Study 2. Teachers tended to ignore the formative assessment activity, as a result of either a perceived or actual lack of time during the lesson. Neglecting the formative assessment episode may indicate teachers are placing a lesser value in terms of learning on this particular episode. Importantly, the majority of the lesson plans ended with a formative assessment activity, typically designed as a plenary activity.

'Teacher talk' was the dominant form of teaching strategy, with two thirds of these activities being described as 'teacher explanation' or 'teacher discussion'. Just under half of the lesson plans involved either a group discussion, group work or paired learning, involving students talking to, or with, other students. The importance of the learner interacting with social settings has been discussed at length (Trebell 2007; Hanks 1991; Eraut 2000).

Lesson episodes tended to be 'teacher-led' rather than 'student-led' (Eraut 2000); that is, activities were 'managed' or 'controlled' by the teacher, suggesting a traditional approach to teaching and learning, with characteristics of a behaviourist paradigm. The teacher dictates what knowledge the learners will learn, in what order they will learn it, how it is to be learnt and, indeed, how it will be demonstrated. When learning outcomes become predictable and measurable (Kennedy 2007) and managed by the teacher, the role of the student in the process of demonstrating their learning is questionable.

Demonstrations were a common activity used by technology teachers and tended to take up a large proportion of the lesson duration. Developing skilled knowledge can often rely on demonstrations, a teaching strategy embedded in constructivist learning. In plan form, the teachers provided no evidence of the knowledge, skills and understanding relating to the particular demonstration; for example, teachers would simply state 'demonstrate drilling'. The demonstrations observed in Study 2 tended to be one-dimensional in relation to learning, in as much as they focused on the procedural knowledge involved in the activity. Whilst this was important, teachers tended to omit any reference to other forms of learning, such as conceptual or technical knowledge. Thus, the demonstrations did not provide the full potential of learning opportunities. Petrina (2007) contends that demonstrations must involve more than how to perform the task, with the teacher modelling what they know and the level of skills and safe practice attained. The learning outcomes associated with demonstrations involved step-by-step instructions on how to achieve the task.

Theme 5: The role of learning outcomes

Identifying possible learning outcomes was not a common or standardised activity in the 'pre-active' phase. Teachers often identified a single learning outcome per lesson, generally relating to the final outcome of the lesson or series of lessons, presumably demonstrating a culmination of learning progress, rather than a series of learning outcomes relating to the intended learning activities or episodes planned into the lesson.

The majority of learning demonstrated in the lessons could be classified as either knowledge-based or practical skills, producing learning outcomes that were either written, sketched/drawn or produced from a practical activity. Study 1 provided evidence of a narrow range of forms of technology learning outcomes. In this regard, there have been several calls for an increase in the variety of learning outcomes in order to represent more fully the range of learning evident in schools (Daugherty et al. 2011; QCDA 2010).

How This Might Be Used to Improve Teaching and Learning

The next section explores the implications of my research on curriculum planning for effective learning in technology and locates my research findings within classroom practices.

- The nature of technology and curriculum planning

Whilst the 'dominant' planning process may be a useful approach when teaching student teachers to plan, as a process it significantly conflicts with effective technology teaching practices in the classroom. The nature of technological activity requires knowledge, skills and understanding on a 'need-to-know' basis (Kimbell et al. 1991). Professional practice in design allows the task or brief to dictate both the most appropriate processes required and the necessary knowledge and skills needed to progress to an effective solution. The teacher needs to plan the learning environment to support, not dictate, the development of the essential knowledge, skills, understanding and processes or key concepts. Furthermore, technology is holistic by nature and requires the student to access the interconnections and interactions between the various stages in the design process, and, as such, the dominant approach to planning cannot support the fundamental requirements of learning progression in technology learning activities.

Such an approach, which develops knowledge and skills on a 'need-to-know' basis, places an emphasis on teaching students a process that involves identifying how and when knowledge is required and not on the knowledge students may one day need (Owen-Jackson and Steeg 2007). Unlike other subject disciplines, this 'just-in-time' learning makes the definition of any specific knowledge boundary difficult (Martin 2011) whilst creating a subject that is unique both in terms of teaching and learning (Middleton 2008; Barlex and Pitt 2000; Kimbell 1997). This emphasis requires a clear view of the role of knowledge in technology teaching and learning and will have implications for the planning, as well as the acquisition of knowledge, through suitable learning activities. 'Just-in-time' learning can be difficult to manage, and the de facto situation, in England at least, is that there is a large body of conceptual knowledge that has to be learned and then deployed through procedural knowledge in designing and making.

Another consideration involves the practical aspects of technology which clearly provides opportunities for the performance evidence of learning (Kimbell 1994). A strong focus on 'doing' and the procedural skills associated with the practical activity, in terms of planning, teaching, learning and assessment, complements the formal dominant planning model and produces manageable learning outcomes through practical learning activities. In situations where the dominant model is used to plan practical learning activities, learning outcomes tend to be performance-based, and learning tends to be conceived in terms of performances. In such situations more metacognitive skills, such as creativity and reflection, are often degraded.

Progressing students' learning in technology requires the interplay or application of knowledge, skills, understanding and processes (Moreland et al. 2008), and learning outcomes need to demonstrate this interplay and application. Learning outcomes need to involve a multidimensional approach to learning or demonstrated interplay between, for instance, knowledge and skill. Planning for learning outcomes that demonstrated 'knowing why' could provide evidence of interplay between two or more learning forms. By considering and planning for learning outcomes that demonstrate different knowledge forms or strands, isolated learning outcomes could be reduced, resulting in a more realistic, authentic learning outcome (Love 2013).

- Curriculum planning models and approaches

Calderhead (1996), like Clark and Peterson (1986) before him, argues that experienced teachers rarely follow a rational model of planning rather a 'creative, interactive, problem-finding and problem-solving process of planning' (1996: 15). In reality, the dominant planning model tends to contrast with more experienced teachers' approaches to planning. John (1991) concludes that teachers' planning processes involve 'looking for good ideas and then for ways of translating these ideas into workable classroom activities' (John 1991: 306).

Several alternative and adapted planning approaches are present in the current literature, which are particularly pertinent to technology education. Borko et al. (1988) suggest that patterns of planning are very much related to a teacher's subject specialism. Although teachers' approaches to planning have generic outlines, a number of individual variations exist which are very much dependent upon the teacher's view of the teaching situation and their own beliefs, values, attitudes and concerns (Tsui 2002). In particular, a teacher's own perception of their subject had a strong influence on the formation of their ideas about planning (John 1991). Tsui (2002) contends that teachers plan in a way that suits their own personal style and the nature of the subject they teach.

The 'naturalistic' or 'organic' model, based on the work of Stenhouse (1975) and Egan (1992, 1997), was developed from the apparent conflict between the need to carefully specify learning intentions and the dynamic nature of classrooms and was an attempt to emulate a realistic planning process based on the 'natural' interactions in a classroom. Naturalistic planning involves starting with activities and the ideas that flow from them before assigning learning objectives (John 2006). Although lacking detail in terms of pedagogical requirements and consideration, this model does resonate with Perkins et al.'s (2000) notion of 'learning in the wild', when learning settings are recognised as 'messy and complex' (Carr 2008: 36). Perkins and Salomon (1992) argue for the need for learners to experience more 'natural' learning environments, with teachers' planning procedures supporting this notion.

The 'interactional method' of planning, another alternative to the dominant model, stresses the interactive nature of learning and, therefore, learning objectives (Brady 1995; Bell and Lofoe 1998). Whilst the 'interaction' model specifies the same design elements as the linear objective model, the 'interactional method' planning process can begin with any of the elements. Based on this model, all curriculum elements interact with each other during the design/planning process, and the design of one element will influence and possibly change the design decisions for other elements. For example, method might be specified first but altered later as a result of an assessment decision. From a practical perspective, this model makes it possible to specify learning objectives after all other elements have been decided (Bell and Lofoe 1998).

The 'articulated curriculum' (Hussey and Smith 2003: 360) provides a similar approach to the 'interactional model', where the respective elements exist in a state of mutual interaction and influence. Alexander (2000) compares this 'articulated curriculum' approach to planning to the structure of a musical performance, where the composition is analogous to the lesson plan and the performance shifts according to interpretation and improvisation. This 'responsive' approach to planning requires the teacher to be vigilant of the learning progression within the class and respond accordingly and is synonymous with the formative assessment principles of 'feedback' (Ramaprasad 1983) (see p. 110). Biggs' (1999) notion of constructive alignment also supports this way of approaching planning for teaching and learning.

- Planning models and approaches and creativity

By providing or developing alternative teaching and learning frameworks for technology pedagogy, teachers develop a range of approaches to planning processes that may be better suited or supportive of the range of intended learning experience needed to progress in technology. Clark and Yinger (1987) claim that teachers use a number of different types of planning approaches, both pre-teaching and during teaching activities, suggesting that teachers do require a 'repertoire' of planning approaches for different requirements, including different pedagogical approaches. Teachers of technology need to consider the implications of their current approaches to planning on the teaching and learning of their subject. As with initial teacher education contexts, planning for teaching and learning provides departments an effective framework to discuss, develop and address the requirements of the subject in a strategic, focused way.

Creativity involves the extended abstract outcomes of learning (Biggs 1999, 2003), such as hypothesising, reflecting, generating ideas, applying knowledge in new contexts or domains and working with problems that do not have unique solutions. As Torrance (2007) discusses in relation to creativity, learning outcomes add value to the information given and are not just the replication of information. Thus, being creative involves complex or multidimensional learning in the sense that it is not about learning something determinate and that, unlike more controlled learning, the outcomes may not be predicted in advance. This is not to say that teachers cannot influence the development of creativity in their students. Teachers can create conditions under which complex learning is more likely to happen and can make general predictions about creativity amongst groups of students. However, complex learning can neither be planned for nor measured in the way that we treat less complex learning (Knight 2002).

The current education context does not allow creativity to flourish (Hallgarten 2014). In relation to this study's findings, there were no examples where students were being asked to either apply or demonstrate their learning in a new context or indeed where creativity was a focus of the learning process. When the focus is firmly on the production of prespecified learning outcomes, which satisfy the monitoring, auditing and reporting requirements of the school, teachers are sensible to ensure they deliver on time and to the necessary quality for each student (Nicholl and McLellan 2009). Technology learning outcomes need to be considered holistically in relation to the learning journey, and teachers need to consider positioning the learning outcomes at relevant points in the learning journey in order to capture authentic learning.

What Else Would Be Good to Know, and How Could Teachers of This Find Out?

Various models supporting technology education have been developed, emphasising and/or reinforcing certain aspects or concepts of the subject and potentially providing the focus for a variety of planning processes.

Morgan et al.'s (2013: 4) approach involves the notion of 'a technology toolbox'. This approach splits technology into four groups: design, technology, critique and data. A series of key concepts and principles is associated with each group and aims to provide a coherent curriculum for technology, which involves an integrated understanding of the key concepts across all material areas. Barlex and Rutland (2004) introduced the 'design decisions pentagon', a conceptual model designed to develop insights into the requirements of teaching designing. The model involved five conceptual considerations: conceptual, marketing, technical, constructional and aesthetic (Rutland 2009; Barlex and Rutland 2004). Moreland's (2008) primary planning tool focused on the multidimensionality of technology, providing teachers with the opportunity to consider conceptual learning outcomes, procedural learning outcomes, societal learning outcomes and technical learning outcomes during the planning process. The Nuffield project with its resource and capability tasks allows teachers to design learning activities that encouraged creativity through designing and making (Barak and Haker 2011). Similarly, the Electronics in Schools Strategy offers eight areas of study that potentially provide an approach to planning that allows teachers to effectively construct a wide variety of different learning activities and cover the varied requirements of technology learning (https://www.data.org.uk/for-education/digital-dt/designing-in-digital-dt/starting-points-for-designing/).

It was encouraging that two of the seven schools planned to develop the idea of a learning journey concept map. One school was intending to use it with their newly qualified teachers to support planning for progression, whilst another school wanted to modify the concept to be used in the classroom with their students. Initially, the school was to use the map with their less-able students to help them identify and provide a clear representation of their learning progress in order to build self-confidence and develop motivation for learning.

ITE is the ideal place to consider planning. Through a process of creation and collaborative planning, 'novice' teachers and their mentors could explore and identify what may or may not be effective within a subject-specific context, and lesson planning proformas could be used as learning aids for initiating discussion around effective technology learning and teaching.

As this research study has demonstrated, the planning process is an integral part of the teaching-learning process and one that needs further exploration in relation to current teaching-learning processes.

References

Alexander, R. (2000). *Culture and pedagogy*. Oxford: Blackwell.

Barak, M. (2011). Fostering learning in the engineering and technology class. In M. Barak & M. Hacker (Eds.), *Fostering human development through engineering and technology education. Reviewing the past twenty years* (pp. 35–53). Rotterdam: Sense Publishers.

Barlex, D., & Pitt, J. (2000). *Interaction: The relationship between science and design and technology in the secondary curriculum*. London: Engineering Council.

Barlex, D., and Rutland, M. (2004). '*Developing trainee teacher's ability to teach designing within secondary school design and technology in England*'. PATT 14 International Conference Proceedings. Albuquerque, USA, March 2004.

Bell, M., & Lofoe, G. (1998). Curriculum design for flexible delivery – Massaging the model. In R. Corderoy (Ed.), *Flexibility: The next wave*. Wollongong: Australian Society for Computers in Tertiary Education.

Biggs, J. (1999). *Teaching for quality learning at university*. Buckingham: SRHE and Open University Press.

Biggs, J. (2003). *Aligning teaching and assessment to curriculum objectives*. LTSN Generic Centre: Imaginative Curriculum Project.

Borko, H., Livingston, C., McCaleb, J., & Mauro, L. (1988). Student teachers' planning and post-lesson reflections: Patterns and implications for teacher preparation. In J. Calderhead (Ed.), *Teachers' professional learning*. Lewes: Falmer Press.

Brady, L. (1995). *Curriculum development*. Wollongong: Prentice Hall.

Calderhead, J. (1984). *Teachers' classroom decision making*. London: Holt, Rinehart and Winston.

Calderhead, J. (1996). Teachers: Beliefs and knowledge. In D. A. Berliner (Ed.), *Handbook of Educational Psychology*. New York: Macmillian.

Carr, M. (2008). Can assessment unlock and open the doors to resourcefulness and agency? In S. Swaffield (Ed.), *Unlocking assessment* (pp. 36–54). Abingdon: Routledge.

Clandinin, D. (1985). Personal practical knowledge: A study of teachers' classroom images. *Curriculum Inquiry, 15*(4), 361–385.

Clandinin, D. (1986). Classroom practice: Teachers images of action. *Journal of Curriculum Studies, 19*(2), 195–200.

Clark, C., & Lampert, M. (1986). The study of teacher thinking: Implications for teacher education. *Journal of Teacher Education, 37*(5), 27–31.

Clark, C. M., & Peterson, P. L. (1986). Teachers' though processes. In M. C. Wittrock (Ed.), *Handbook of research on teaching* (3rd ed., pp. 255–296). New York: Macmillian.

Clark, C. M., & Yinger, R. J. (1987). Teacher planning. In J. Caldehead (Ed.), *Exploring teachers' thinking* (pp. 84–103). London: Cassell.

Datta, L. (2001). The wheelbarrow, the mosaic and the double helix: Challenges and strategies for successfully carrying out mixed methods evaluation [online], *Evaluation Journal of*

Australasia, 1(2), 33–40. Available at: http://search.informit.com.au/documentSummary;dn=456685309173089;res=IELBUS>ISSN:1035-719X.%5B. Accessed 14 Mar 15%5D.
Daugherty, R. B., Black, P., Ecclestone, K., James, M., & Newton, P. (2011). The assessment of significant learning outcomes'. In R. Berry & B. Adamson (Eds.), *Assessment reform in education* (pp. 165–183). New York: Springer.
Egan, K. (1992). *Imagination in teaching and learning*. Chicago: University of Chicago Press.
Egan, K. (1997). *The educated mind: How cognitive tools shape our understanding*. Chicago: University of Chicago.
Elliott, J. (2001). Making evidence-based practice educational. *British Educational Research Journal, 27*(5), 555–574.
Eraut, M. (2000). Non-formal learning and tacit knowledge in professional work. *British Journal of Educational Psychology, 70*, 113–136.
Greene, J. (2008). Is mixed methods social inquiry a distinct methodology? *Journal of Mixed Methods Research, 2*(1), 7–22.
Guest, G. (2012). *Applied thematic analysis*. Social Research Solutions (pp. 1–320).
Hagger, H., & McIntrye, D. (2006). *Learning teaching from Tearchers: Realising the potential of school-based teacher education*. Maidenhead: Open University Press.
Hallgarten, J. B. (Ed.). (2014, November 7). *Licensed to create. Ten essays on improving teacher quality*. London: RSA.
Hanks, W. (1991). Foreword. In J. Lave & E. Wenger (Eds.), *Situated learning: Legitimate peripheral participation*. Cambridge: Cambridge University Press.
Hewitt, D. (2008). *Understanding effective learning: Strategies for the classroom*. Maidenhead: Open University Press.
Hussey, T., & Smith, P. (2003). The uses of learning outcomes. *Teaching in Higher Education, 8*(3), 357–368.
Illeris, K. (2009). In K. Illeris (Ed.), *Contemporary theories of learning: Learning theorists...in their own words*. London: Routledge.
John, P. (1991). A qualitative study of British student teachers' lesson planning perspectives. *Journal of Education for Teaching, 17*(3), 301–318.
John, P. (2006). Lesson planning and the student teacher: Re-thinking the dominant model. *Journal of Curriculum Studies, 38*(4), 483–498.
Kennedy, D. (2007). *Writing and using learning outcomes, a practical guide*. Cork: Quality Promotion Unit, UCC.
Kimbell, R. (1994). Progression in learning and the assessment of children's attainment. In D. Layton (Ed.), *Innovations in science and technology education*. Paris: UNESCO.
Kimbell, R. (1997). *Assessing technology, international trends in curriculum and assessment*. Buckingham: Open University Press.
Kimbell, R., Stables, K., Wheeler, T., Wosniak, A., & Kelly, V. (1991). *The assessment of performance in design and technology*. London: SEAC and the Central Office of Information.
Knight, P. T. (2002). *Being a teacher in higher education*. Buckingham: SRHE and Open University Press.
Love, T. (2013). *Theoretical underpinnings towards assessing science pedagogical content knowledge (PCK) of technology educators. PATT 27 conference technology education for the future: A play on sustainability* (Vol. 27, pp. 291–297). Christchurch: University of Canterbury.
Martin, R. (2011). Innovation in symbolic industries: The geography and organisation of knowledge sourcing. *European Planning Studies, 19*, 1183.
McAlpine, L., Weston, C., Beauchamp, J., Wiseman, C., & Beauchamp, C. (1999). Building a metacognitive model of reflection. *Higher Education, 37*, 105–131.
Middleton, H. (2008). *Researching technology education: Methods and techniques*. Rotterdam: Sense Publishers.
Milkova, S. (2014). *Strategies for effective lesson planning*. [online]. Available from Center for Research on Learning and Teaching: crit.umich.edu. Accessed on 26 Nov 2014.

Moreland, J. (2008). Assessment for learning in primary technology classrooms. *Design and Technology Education: An International Journal, 12*(2), 37–48.
Moreland, J., Jones, A., & Barlex, D. (2008). *Design and technology inside the black box – Assessment for learning in the design and technology classroom.* London: GL Assessment.
Morgan, R., Jones, L., & Barlex, D. (2013). New Principles for Design & Technology in the National Curriculum. *Education for Engineering. E4E2013, London, UK.*.
Mutton, T. H., Hagger, H., & Burn, K. (2011). Learning to plan, planning to learn: The developing expertise of beginning teachers. *Teachers and Teaching: Theory and Practice, 17*(4), 399–416.
Nicholl, B., & McLellan, R. (2009). 'This isn't my project [work]. It's…just do it… you just do research'. What student voice reveals about the nature of D&T lesson in English schools and the implications this has on their motivation and learning of complex tasks. In M. J. de Vries (Ed.), *International handbook of research and development in technology education.* Rotterdam: SENSE.
Owen-Jackson, G., & Steeg, T. (2007). The role of technical understanding in design and technology. In D. Barlex (Ed.), *Design and technology – The next generation: A collection of provocative pieces to stimulate reflection and curriculum innovation London, UK.* TEP/Nuffield.
Perkins, D. N., & Salomon, G. (1992). *Transfer of learning. International encyclopedia of education* (2nd ed.). Oxford: Pergamon Press. [online]. Available at: http://www.cdtl.nus.edu.sg/Ideas/iot18.htm. Accessed on 31 Mar 2013.
Perkins, D., Tishman, S., Ritchart, R., Donis, K., & Andrade, A. (2000). Intelligence in the wild: A dispositional view of intellectual traits. *Educational Psychology Review, 12*(3), 269–293.
Peters, R. S. (1965). *'Education as initiation'. Inaugural Lecture*, University of London, Institute of Education.
Petrina, S. (2007). *Advanced teaching methods for the technology classroom.* Hershey: Springer.
QCDA. (2010). *Guidance for senior leaders, Qualification and Curriculum Development Agency, Department of Education, London, UK.*, Coventry.
Ramaprasad, A. (1983). On the definition of feedback. *Behavioural Science, 28*, 110.
Rutland, M. (2009). Art and design and design and technology: Is there creativity in the designing? *Design and Technology Education: An International Journal, 14*(1), 56–67.
Shavelson, R. J., & Stern, P. (1981). Research on teachers' pedagogical thoughts, judgements, decisions, and behaviour. *Review of Educational Research, 51*, 455–498.
Sondergeld, T. B. (2010). Understanding how teachers engage in formative assessment. *Teaching and Learning, 24*(2), 72–86.
Stenhouse, L. (1975). *An introduction to curriculum research and development.* London: Heinemann.
Torrance, H. (2007). Assessment as learning? How the use of explicit learning objectives, assessment criteria and feedback in post-secondary education and training can come to dominate learning. *Assessment in Education: Principles, Policy and Practice, 14*(3), 281–294.
Trebell, D. (2007). *A literature review in search of an appropriate theoretical perspective to frame a study of Designerly activity in secondary design and technology.* Education and international research conference (pp. 91–95). The Design and Technology Association.
Tsui, L. (2002). Fostering critical thinking through effective pedagogy: Evidence from four institutional case studies. *The Journal of Higher Education, 73*(6), 740–763.
Tyler, R. (1949). How can learning experiences be organised for effective Instruction? In *Basic principles of curriculum and instruction.* Chicago: University of Chicago Press.

Chapter 10
Enhancing the Teaching of Problem-Solving in Technology Education

Thomas Delahunty

Abstract Problem-solving skills are a critical element of science, technology, engineering and mathematics (STEM) education, and improving students' capability is a contemporary focus. Traditional research in this area has emphasised processes and heuristic approaches to solving problems while neglecting the early stages such as problem conceptualisation. This chapter will discuss a research study aimed at addressing this knowledge gap in technology education using a novel approach adapted from the field of cognitive neuroscience. Findings are then discussed in the context of enhancing pedagogical practice around the framing of problem-solving tasks on the part of the teacher and the student. This chapter will be of interest to a wide audience of educators in STEM disciplines.

Conceptualising the Issue

One of the most pressing focuses in contemporary STEM education is the development of robust, flexible and fluent problem-solving skills. Any recent review of reports and initiatives on the development of STEM competencies, for example, will espouse the necessity for a graduate workforce with these attributes. Despite these aspirations, concerns have been raised about the current quality of graduates entering the workforce and in particular the lack of epistemically fluent approaches to solving problems. These concerns are not confined to the vocational context of the workforce but are evident in the earlier stages of education as well. For example, McCormick and Davidson (2009) have described the linear and rigid problem-solving skills promoted by the inflexible conception of product-based outcomes within problem-based learning (PBL).

While a great deal of research work has focused on the processes and related heuristics involved in solving problems, little is known about the fundamental

T. Delahunty (✉)
University College Cork, Cork, Ireland
e-mail: thomas.delahunty@ucc.ie

P J. Williams, D. Barlex (eds.), *Explorations in Technology Education Research*, Contemporary Issues in Technology Education,
https://doi.org/10.1007/978-981-13-3010-0_10

processes underlying an individual's ability to conceptualise a problem situation. In reference to instructional design from a teacher's perspective, this is a critical issue as it is posited that the manner in which an individual represents a problem impacts on the subsequent approach and efficacy. Despite this hypothesis and to the best of my knowledge, there is no empirical evidence supporting this within an educational context. Therefore, one of the key goals of this research was to investigate if such an empirical relationship between conceptualisation and approach exists within the context of solving a problem. In addition, an exploration of the cognitive elements that support and constitute a conceptual frame was conducted. This included an analysis of cognitive processes and perceptive elements that may impact on the process of conceptualisation. It is one thing to determine if there is a relationship between conceptualisation and approach; it is also a necessity to understand the elements that support or indeed impair this relationship.

From a teaching perspective, these goals are central to enhancing our understanding of instructional design and implementation within a PBL environment such as that within a technology education classroom. The core focus of the research reported here was to address the gap in knowledge surrounding the conceptualisation of problems utilised for developmental purposes within an educational context and associations with performance. This was a complex undertaking as the conceptualisation of a problem is posited to be an extremely subjective multifaceted process incorporating elements such as past experience, environmental influences, epistemic orientation and cognitive ability among others (Payne et al. 2013; Kitchner 1983). The next section of this chapter will explain the methodology by which the author attempted to answer these queries. As a guide, the research questions which guided the study are provided:

1. Is there a relationship between the nature of the problem conceptualisation and the approach adopted by the individual?
2. What elements influence the manner in which individuals conceptualise convergent problems?
3. Is there a relationship between specific processes, the nature of the conceptual frame and ultimately performance?

Research Approach

In order to investigate these research questions, a novel methodology was designed utilising a pragmatic mixed-methods approach. The pragmatic application within this study avoided any absolute epistemological or ontological assumptions (Feilzer 2010), and this was an important principle given the idiosyncratic nature of an individual's conceptual framing and the lack of empirical data available on the process. The remainder of this section will explain the various methods employed to investigate the process of conceptualisation during problem-solving, and Fig. 10.1 illustrates the complete collection of tools employed. The methods encompassed three areas which will be elaborated on in the following subsections.

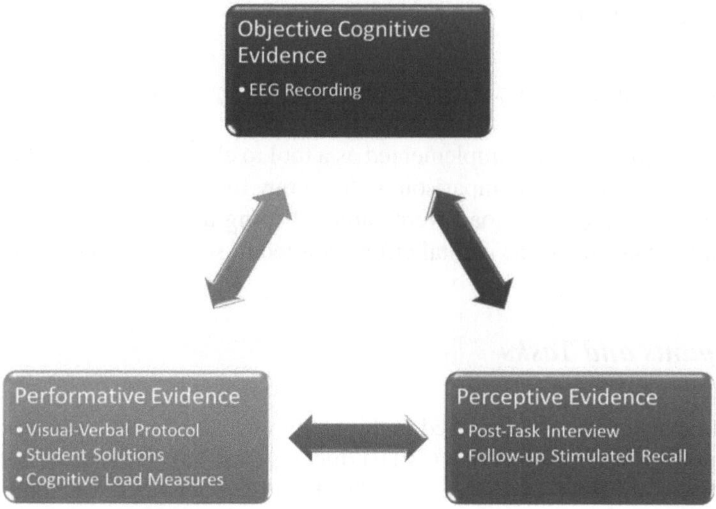

Fig. 10.1 Triangulated methods

Evidence of Cognitive Processing

Given the goals of this research, capturing data relating to cognitive processing was a key consideration. The field of cognitive neuroscience has recently been heralded as an avenue with potential advantages for educational research applications (Arrowsmith-Young 2012). While a number of approaches to gather cognitive evidence, within this field, were considered, it was decided to employ electroencephalography (EEG) as commercial headsets are available, and more importantly this method allows for temporal analysis of the problem-solving episodes, which was a distinct advantage for this study.

Perceptive Evidence

Considering the posited subjectivity of experience associated with the conceptualisation process, it was important to gather evidence from the individuals' perspective. This enhanced the objective cognitive data set by providing rich evidence of students' perceived approach to solving the prescribed tasks. Two post-task interviews were implemented for this purpose: one immediately following the problem-solving session and the other a day after the event. Both interviews followed a semi-structured format.

Performance Data

As shown in Fig. 10.1, performance was captured by multiple methods. Primary performance was calculated using the solutions produced by the participants. Visual-verbal protocol was implemented as a tool to clarify the approaches adopted by students and facilitate comparison with the raw EEG data. Additionally, subjective measures of cognitive load were captured using a semantic differential scale where participants rated the mental effort required to solve each problem.

Participants and Tasks

The participants that were involved in this study were third-year undergraduate initial technology teacher education (ITTE) students ($N = 12$). These participants were chosen based on their experience in STEM problem-solving (the context of the problems), their presumed competencies in a variety of problem-solving strategies and their reflective capacities which would facilitate meaningful discussion in the post-task interviews.

The tasks for this study were chosen to encompass three different categories: mathematical, visuospatial and dual approach. The rationale for this was to capture data relating to different cognitive approaches (mathematical and visuospatial) and also to observe the flexibility of participants' conceptual framing when the task could be solved in different ways. The tasks were taken from a database designed by O'Donoghue and Kooij (2007) based on principles adopted from the Programme for International Student Assessment (PISA). The databases of tasks are detailed in Table 10.1 and were selected for their already validated nature and to control for difficulty (being designed for 15-year-olds originally) which could impact on the conceptualisation process and EEG data.

A sample of the tasks that were used is shown in Fig. 10.2. The tasks are convergent in nature meaning that they have a single correct solution. The key principle in their selection was that the mathematical and visuospatial categories were best solved using a single respective approach. The dual-approach category still had a single correct response, but the approach was more ambiguous – it could be solved in more than one way, but there may be an optimal approach.

Key Findings

This section of the chapter will discuss some of the key findings garnered from the implementation of the previously described methodology. While the original study by Delahunty (2014) gathered a significant data set, there is only limited space within this chapter to delineate the core relevance to teaching and learning within technology education. Therefore, only the data most relevant to this context will be presented.

Table 10.1 Task typology for Delahunty (2014) study

Category	Task	Descriptor
Mathematical	A	Application of theorem
	B	Simple arithmetic
	C	Application of formula
	D	Algebraic
	E	Identification and application of theorem
Dual approach	F	Simple arithmetic/image holding and comparing
	G	Simple arithmetic/image holding and comparing
	H	Application of formula/image holding and comparing
	I	Application of formula/image holding and comparing
	J	Probability/image holding and comparing
Visuospatial	K	Dynamic imagery
	L	Image holding and comparing
	M	Planar rotation
	N	Dynamic imagery
	O	Kinetic imagery

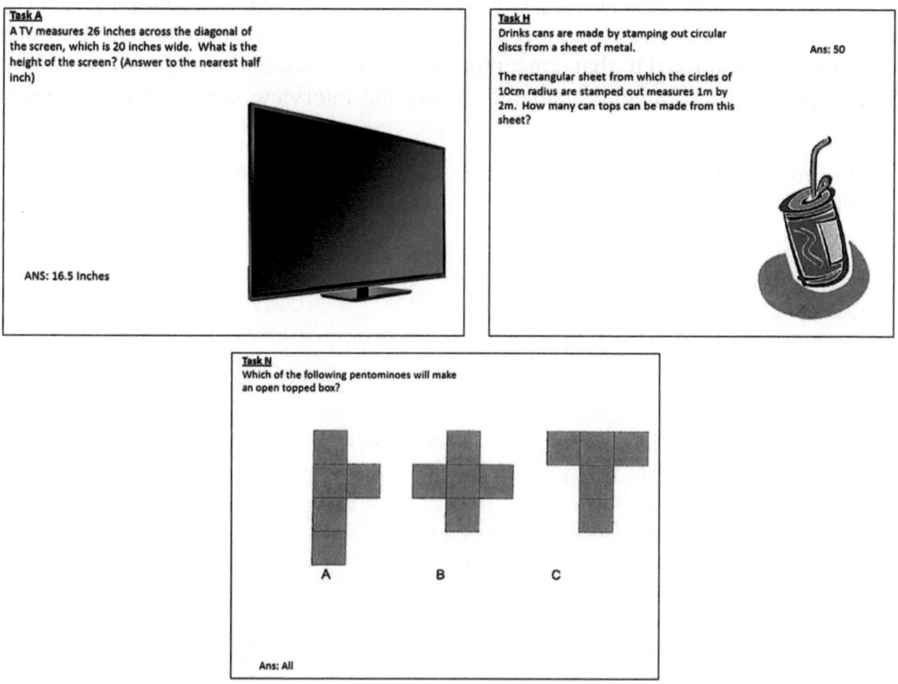

Fig. 10.2 Sample tasks from each category

The Relationship Between Conceptualisation and Approach

Given the lack of empirical data available on the relationship between an individual's conceptualisation and subsequent approach to solving a problem, it is important to firstly consider this data. This relationship was investigated using a combination of cognitive and perceptive data. The key strength of the study lay in the application of an objective measure of cognitive processing by means of EEG. This allowed the researcher to observe, on a cognitive level, whether a relationship existed. A brief note on this process of data analysis is required here. Raw EEG data comes in the form of a complex waveform (Fig. 10.3) and contains a variety of artefacts and "noise". The key requirement at this stage is to remove these contaminants so that a clear picture of cortical cognitive activity can be observed (Delorme and Makeig 2004). This process then allowed the data for each participant to be plotted as topographic maps displaying the pattern of cognitive activity across the cerebral cortex. To observe the existence of any cognitive relationship between conceptualisation and approach, these topographic maps were plotted in temporal stages of the problem-solving episode. A representative sample of these maps is shown in Fig. 10.4.

The data shown in Fig. 10.4 demonstrate a clear pattern. The cognitive activity evident at the beginning of the task does not alter in any significant manner for the remainder of the problem-solving episode. That is to say, once a participant conceptualises a problem, at least from a cognitive perspective, their approach is very much dependent on and fixed to that conceptual frame. This relationship is also supported in the qualitative comments gathered during the interview stages of the research study:

> How you interpret the problem, may be your interpretation of the problem influences what approach you take. (Participant 02)

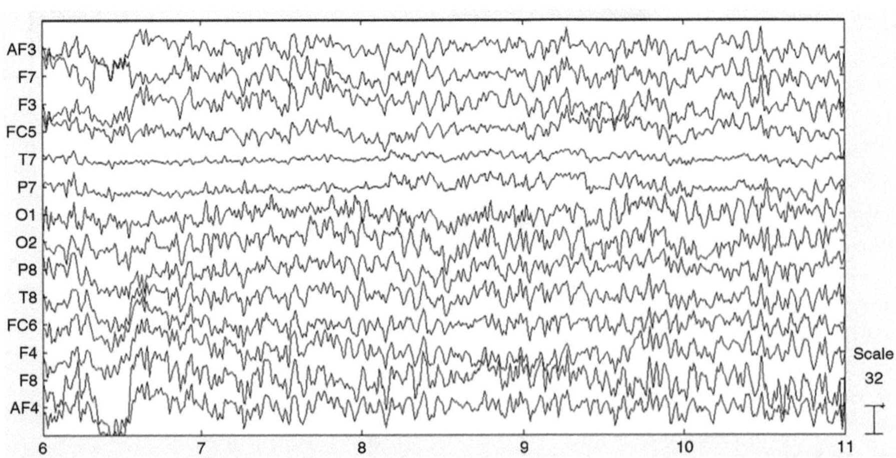

Fig. 10.3 Raw EEG signal

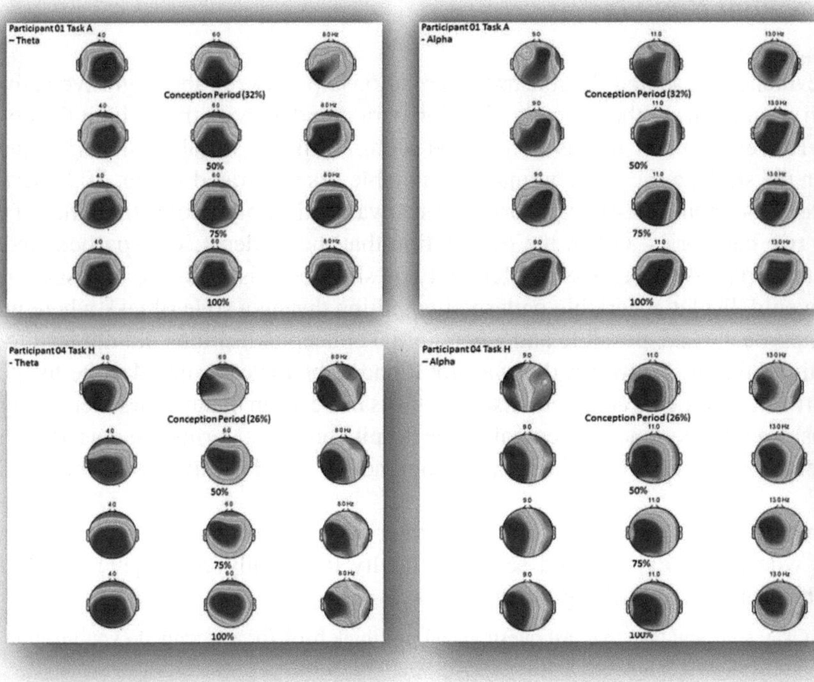

Fig. 10.4 Sample of topographic plots

> I was kind of forming the approach in my head… it was immediate reaction to jump to algebra once I read the question it was I suppose from other questions I might have done before the approach I would have taken. (Participant 02)

The above comments taken from the same participant in both interview sessions indicate a belief in the relationship between conceptualisation and approach:

> R: "How did you conceive the task?"
> Participant: "I thought it was going to be simple because I was going slow there thinking that it was going to be okay, but then as I done my first two steps going eastwards and then southwards, I seen all the different possibilities that I could use to actually get to B."
> R: "So if you were to describe your actual approach to solving the question?"
> Participant: "… just kind of highlighting the line in my head, I can remember." (Participant 07)

Evident in the above participant's comments is a direct relationship between the initial conception indicated by "seeing different possibilities" and the description of the approach. A visual conception would appear to have initiated a visual approach to this task. This data supports the hypothesis that conceptualisation and approach, during problem-solving, are related to each other. This trend appears in the majority of participants' data and interestingly occurs regardless of performance outcome. This is the second key area that is worth exploring in the context of teaching and learning.

Conceptual Rigidity

The evidence presented demonstrates a relationship between the cognitive activity which occurs during the conceptualisation period and activity throughout the entire task. However, when this is considered along with the variable of performance within the study, a second key finding is notable. As illustrated in Fig. 10.5, performance in the non-specific category of tasks was noticeably poorer than that of the other two categories. Given the expectation that the students, who participated in this study, possess a flexible problem-solving skill set, it is plausible to assume that they would display a variable pattern of cognition during a style of task where multiple approaches are possible and where initial conceptualisations did not lead to an effective solution. However, this was not found to be the case, and despite the poor performance, students conceptualised the tasks in a certain manner and then engaged the tasks with this same conceptual frame despite progress during the episode. They then either accepted this as the suitable approach or could not think of an alternative one.

This finding is also supported in the interview commentary by several participants where they experienced severe difficulty in initially conceiving alternative possibilities for approaching the problems:

R: Okay so if you take that last question and just think back for a second, do you imagine there was any other way you could have approached it?

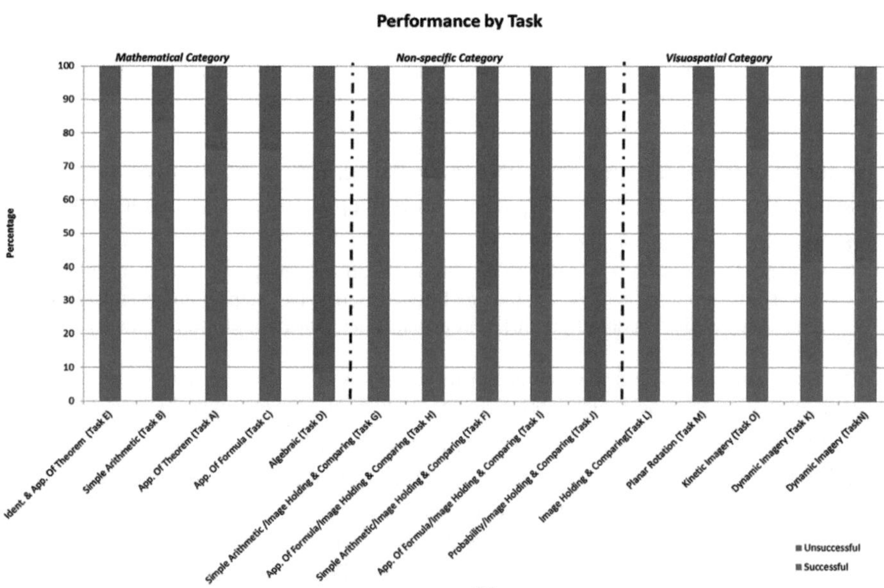

Fig. 10.5 Performance in problem-solving tasks

Participant 01: … for me no, that's the way I would have answered it unless I had seen another way maybe I would have followed that way but for me that was the only way I would have approached it.
R: Okay so really you didn't think of any other approach you decided on that one and you were dead set?
Participant 01: Yes yeah.

Some transcripts also demonstrated some evidence of a lack of confidence in approaching tasks in alternative ways:

R: Was there any other way you could have approached it?
Participant 06: … well you could have drew it out graphically. You knew straight away that you know 2 cm is half of 4 so you knew that you were going to get 8. You know so it was only a matter of drawing it out.
R.A: So why did you choose the approach that you did?
Participant 06: Because I don't know when you do it mathematically that at least you know that when you get your final answer you know it's right like.
R: Okay.
Participant 06: So I don't know why. I don't know why I done it that way.

There seems to be a lack of metacognitive regulation and understanding evident in these participants' approaches in the immediate post-task interview. This of course is one of the hallmark attributes of expert adaptive problem-solvers, so it is concerning that these students display little of this attribute. While this is an important point to consider as it concerns potential impact on participants' future practice, it is also important to consider some of the potential reasons for this pattern of conceptual rigidity.

Influence of Past School Experience

As discussed in the work of Hahn and Chater (1998), the role of memory and past experience in problem-solving and reasoning scenarios is unquestionable. Often, participants faced with a problem will compare the current situation to a past experience during the conceptualisation of the situation. Reliance on past experience and in particular experience informed by the participants' past schooling was a common trend found in the data:

> Yeah maybe from school you kind of go into a bank of memory rather than maybe trying to work it out… (Participant 01)

> …I suppose it was…driven into me from that stage (school)… (Participant 02)

> …you'd be doing a lot of maths problems even in secondary school… (Participant 08)

The constant reference to school experience and in particular "math"-type problem-solving experience was a strong trend in the overall data. Given the rigidity discussed in the previous section, it is plausible to posit that elements of conditioned behaviour were exhibited in solving the 15 prescribed tasks and that this conditioned behaviour stemmed from a conceptual frame primarily informed by past experience of schooling.

The findings present a number of key points for consideration. Firstly, the evidence suggests a relationship between conceptualisation and approach when solving these convergent tasks. While the ultimate goal is to provide students with a skill set to engage in complex ill-defined problems, convergent tasks are extremely important for developing these foundational competencies. These future teachers also exhibited a tendency to conceptual rigidity in solving these tasks, and this was informed, at least in part, by their past experiences in school. This has implications for their future practice but also raises questions around educational practices at the earlier school levels.

Improving Problem-Based Teaching and Learning

It is important to note that conceptualisation is not the same as problem representation. Representation refers to the summary of an individual's understanding of a problem, whereas conceptualisation is the more holistic process that leads and makes use of subsequent representations (Delahunty et al. 2018). Therefore, conceptualisation is a much more multifaceted process encompassing cognition, personality traits and situated factors. As such, this section, with recommendations for improving teaching and learning, is not intended to be a panacea but more so a set of recommendations based on evolving and continuing research in the area. The data gathered demonstrates a rigidity in the approach to conceptualising convergent tasks that are ubiquitous in education curricula and often used for developmental purposes. This advances our knowledge of problem-solving performance as it has been previously shown that students often adopt procedural and rigid approaches in the face of over-restrictive curricula (McCormick and Davidson 2009). The research and data described in this chapter highlight the exact nature of this rigidity occurring during the conceptualisation stage of problem-solving tasks. It is not my goal to analyse potential causality here but to state the existence of this phenomenon and explore potential recommendations for teaching. As the data here shows, these conceptual rigidities that have been (at least somewhat) cultivated in school are pervasive and affect students much further into their academic career.

Highlighting recommendations is not a straightforward endeavour given the complexities involved in problem-solving as a human activity. So, the key question is where to start?, and this may begin by considering our overarching goals within an educational context. It is well documented that factors such as summative examinations influence both the nature of teaching and learning often leading to surface approaches being adopted (Entwistle 2000). These approaches tend to focus on memorisation and reproducibility. Now, also consider that well-defined convergent tasks are the most common style found in curricula (Jonassen et al. 2006). These styles of tasks lend themselves to the development of domain-specific schema and knowledge and are the underpinning principle of instructional design theories such as cognitive load (Sweller 1988; Sweller et al. 2011). However, while cognitive load theory advocates the use of explicit instruction practices to develop novices' skills prior to more advanced learning and problem-solving, it makes the specious assumption that the development of domain schema and knowledge is the optimal and often only instructional goal. Indeed, the evidence gathered by Delahunty (2014) suggests the negative consequences of over-adherence to such a goal manifesting as students' inability to flexibly conceptualise and reconceptualise problems. Cognitive load theory has provided many instructional approaches that are particularly effective in developing domain-specific schema, but as Kalyuga and Singh (2016) discuss, a much needed rethinking of its boundaries is required. It is clear that instructional goals are not the same in every field of study, and indeed an interesting framework was postulated by Kalyuga and Singh (2016) that encapsulates instructional goals in three hierarchical levels. I have summarised these levels in Fig. 10.6.

It is arguable that with current curricular focuses on examination performance teachers and students are primarily targeting a level 2 focus. That is not to say that they have attained competence in level 1 goals, but rather they may be ignoring them. Refocusing on problem conceptualisation, it is clear that all levels of the framework are critical but specific to the research presented in this chapter; level 1 is a key area in this regard. In order to target this pre-instruction goal level, it is clear that a more sociocognitive lens must be adopted. In exploring recommendations for teaching and learning, I will adopt the perspective of the *opportunity-propensity* framework (Byrnes and Miller 2007) which focuses on creating the opportunity for students to learn and encouraging the propensity to engage with such learning opportunities.

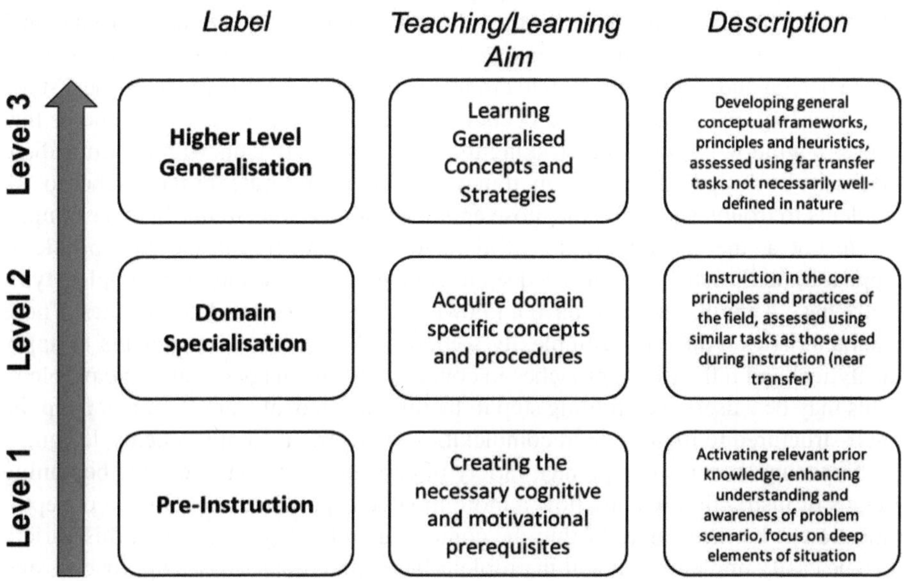

Fig. 10.6 Instructional goals. (Adapted from Kalyuga and Singh 2016)

Designing/Selecting the Epistemic Artefacts to Provide Opportunity

It is clear that the primary role of the classroom teacher is to provide and cultivate the opportunity for their students to learn. This includes designing or selecting appropriate tasks and activities as well as the pedagogical approaches of questioning, assessment and more. Firstly, consider that the majority of problems that are encountered, by students, in most textbooks are convergent or well-defined in nature (Jonassen et al. 2006). They are typically suited to a particular approach and have a single correct solution. This would align with the approaches alluded to in student commentary within the current chapter but also presents us as educators with a serious issue for consideration. In technology education, problem-based and design-learning scenarios are heralded as key instructional approaches. Here, engagement with ill-defined problems in a creative manner is seen as the ultimate goal. Scaffolding novice students to reach these higher levels of problem-solving expertise is a core responsibility of the teacher. Problem types must therefore be seen as more than a dichotomy between convergent (well-defined) and divergent (ill-defined) styles. A useful framework which could be used to guide the design and/or selection of tasks is the problem typology of Jonassen (2011). This typology encompasses a continuum ranging from well-defined to ill-defined problem types and includes styles such as algorithms, story problems, design problems and dilemmas among others. This is illustrated in Fig. 10.7.

Being aware and implementing a greater variety of task styles is one method to cultivate opportunities where flexible problem conceptualisation is promoted and developed. To give an example for thought, consider the goal of enhancing students' abilities in electrical circuit designs. Perhaps, our goal is to improve students' ability to design and implement simple circuits using sensors to solve an "if-then"-type problem (e.g. when a high wind occurs, we want a sensor to activate a motor that closes a window). Rather than tasking the more novice students with the immediate necessity of designing this system, it may be more appropriate for the teacher to ask students to troubleshoot and diagnose an issue with a given circuit. In this example, while not at the capstone of the ill-defined spectrum, troubleshooting problems appear to be ill-defined, require a deep understanding of the circuit principles, typically have a single fault and have a known set of solutions and approaches. These characteristics reduce the complexity somewhat but still require students to apply analytical and reflective approaches in conceptualising an approach to the problem. This may be a useful scaffolding step in facilitating students' development from the well-structured to ill-structured complexities of electrical circuit systems design.

Exposing students to a greater variety of epistemic artefacts from the beginning stages of instruction would allow a more flexible experience of problem conceptualisation to be inculcated. Within the context of technology education, this variety can be easily implemented within problem-based and project-based approaches and not only provide a more encompassing experience for individuals but also facilitate the opportunity for differential sociocognitive interaction among students. At all

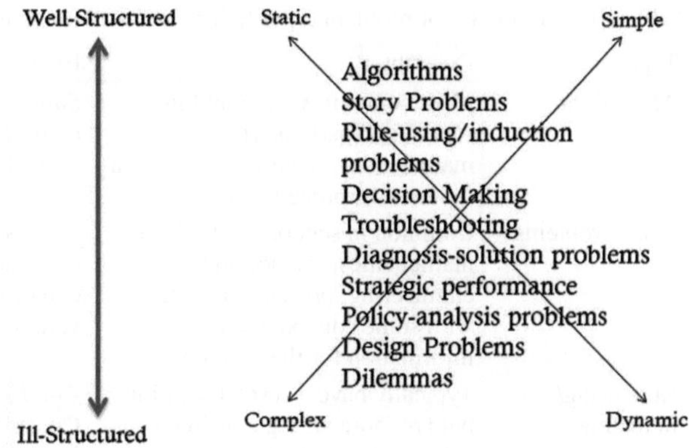

Fig. 10.7 Problem typology. (Jonassen 2011)

times, it is useful to take a macro perspective of the overall problem-based educational aim and ask "Is there an opportunity to select/design an alternative task or form of a task to better facilitate students in developing these flexible and adaptive problem solving behaviours?" As discussed by Markauskaite and Goodyear (2017) in their work on developing epistemic fluency, by exposing students to a greater variety of instructional interventions, they will inevitably develop a greater repertoire of cognitive strategies and conceptual abilities.

Unfortunately, I cannot go into intricate detail of each problem type in Jonassen's taxonomy; however, I have provided a brief overview with descriptions of the task types in Table 10.2. In addition, I have also summarised some example types and some of the associated problem-solving processes involved in each category of task. For a full-detailed description of each type, I direct the reader to Jonassen (2011).

We have looked at the idea of utilising a greater variety of task types beginning at the pre-instruction level. These form the epistemic artefacts or objects of inquiry that students can engage with to stimulate flexible conceptualisation and problem-solving approaches. However, the artefact of learning is not the only concept that requires attention. We also need to consider the social frame within which this activity will occur. This is particularly important in the context of the research data on conceptualisation, presented here, as past social experiences are a source cited in student commentaries. In the next section, I will explore these ideas that focus on practices to increase students' propensity to engage with the learning opportunity.

Encouraging Introspection to Increase Propensity

While reflective capacities have been widely acknowledged as important in a variety of fields, it is in the field of professional education that they appear quite explicitly. Markauskaite and Goodyear (2017) discuss the use of epistemic games as a key to achieving an inquiry-based environment that cultivates learning processes

Table 10.2 Summary of problem types (adapted from Jonassen 2011)

Type	Descriptor	Examples and processes
Algorithms	Most common style found in school and particularly mathematics, requires the use of a finite set of procedures	*Simple calculations*: Number comprehension, production, calculating variables
Story problems	Common in school textbooks in mathematics, science and engineering; presents students with some context and key information for the problem	*Textbook story problems*: Representing unknowns, translating relationships into equations, solving equations, verifying values
Rule using/ induction	Typically have a correct solution but are more ambiguous in the solution method which may be applied	*Conducting a library search*: Constructing effective search arguments, implementing a strategy, evaluating the usefulness of the information
Decision-making	Require the solver to select a solution from a set of alternatives, typical type encountered in everyday reasoning	*Selecting an appropriate school for your child*: Assigning weights to cues, evaluating alternatives, interpreting factors associated with decision
Troubleshooting	Very common in everyday contexts, typically associated with technical professions	*Finding a fault in a mechanical system*: Problem space construction, generation of causes, testing and evaluation
Diagnosis solution	Similar to troubleshooting in that it involves identifying a fault but less well-defined in terms of solution appropriateness	*Patient diagnosis*: Examination and analysis, hypothesis generation and testing, consideration of situated factors (e.g. insurance company policies), evaluation
Strategic performance	Occur in dynamic, changing and ambiguous situations; occur in real-time; are ill-structured and may contain competing goals	*Battlefield commander*: Select or devise a set of tactical strategies, adjustment of strategies in real time depending on progress, contextually constrained
Policy analysis problems	Complex problems or issues on which multiple perspectives and positions exist, ill-structured	*How do we provide affordable housing for people living near the poverty line?*: Defining the problem, gathering evidence, generating alternatives, selecting and applying criteria, envisaging outcomes, balancing trade-offs, deciding and communicating
Design problems	Ill-structured; have vague or unclear goals and multiple and uncertain constraints; have multiple solutions and solution paths	*Designing a new product for a customer*: Conceptualising and defining problem space, applying prior knowledge, analysing constraints, modelling, decision-making, communication
Dilemmas	Possibly the most ill-structured and dynamic problem type; often no clear solution that will be acceptable by all; situation is so complex and unpredictable that no optimal solution can ever be known	*Debating on ethical issues such as same-sex marriage and abortion*: Defining and conceptualising problem spaces, conducting research, managing conflict, articulation and defence of solution

focused on knowing as opposed to knowledge. These styles of environment function primarily through discourse. They premise this approach on the teacher using both social and epistemic scripts to guide the students during the epistemic game. The epistemic scripts refer to guidance regarding the solving of a particular problem and involve questions on key principles and procedures involved in solving a problem within a particular domain. The social scripts are related to guiding students in terms of their interaction and discourse with each other mediated by the activity (epistemic artefact) assigned. It is arguable that the style of educational provision highlighted in the student commentaries in this chapter represents a focus on epistemic scripts. In order to develop more flexible conceptualisation skills in students, it is necessary to provide an environment where problem-solving approaches can be explored, communicated, critiqued and evolved in a collaborative and constructive manner. This is the rationale for focusing on the epistemic scripts that teachers can use to support such an environment.

Perhaps the most simple and pragmatic manner to aid students in reflecting upon their problem conceptualisation is to ask them to introspect on the process. Jakel and Schreiber (2013) discuss research that supports this approach demonstrating that students who are asked to introspect and reflect on their approaches demonstrate greater abilities to transfer to more novel tasks. Taking this a little further, a collaborative activity can be created where students may be assigned different problems or different representations of the same problem in order to elicit alternative approaches. Following their individual solution procedures, they can then be asked to present and critique their approaches. A template can be provided for this critique to help students interact meaningfully with each other and may be in the form of a question or a template for written response. Of course, the number of prompts and scripts that a teacher can employ is quite large and varied and would depend entirely on the instructional activity. Considering making the activity a social one would aid in creating the necessary conditions advocated by level 1 pre-instruction goals (Fig. 10.6). It is crucial for adaptive and flexible conceptualisation skills to develop that teachers embrace a deeper understanding and varied the use of epistemic artefacts and social scripts.

Summary

This chapter has presented evidence that highlights the importance of focusing on the conceptualisation of educational problems. It has done so through the lens of cognitive psychology and by exploring the neurocognitive basis of cognitive processing and the phenomenological parameters of individual approach in a series of prescribed tasks typical of educational interventions. What is apparent is that the overemphasis on specific problem-solving approaches and an over adherence to a specific task design archetype can limit students' opportunity and propensity to develop flexible and adaptive problem-solving skills. In adapting the findings presented here to potential classroom practice, I have drawn on the theoretical problem task typology of

Jonassen (2011), and the reader is directed here as a useful source to gain greater insights into the details of each task type and supporting discussions.

Utilising a greater variety of task types, thus exposing students to a greater variety of conceptual approaches, is likely to have positive effects; however, this is just one element to consider. The notion of conceptualisation advocated in this chapter is a different idea to that of problem representation. It is likely that the result of the conceptualisation process leads to a representation for a problem instance (Delahunty et al. 2018), but the two are inextricably linked. Unfortunately, I do not have the opportunity to delve into this distinction here, but my goal in raising this point is to highlight the area of representation for classroom practice. Students must be equipped with the prerequisite skills, such as sketching capacities, in order to support effective problem representation. In addition, it may be useful for the educator to ask themselves about the alternative approaches to representing some of the problems that form a core focus of their lessons. Showing students alternative methods of representing problems should facilitate them in conceptualising problems in different manners and open up the opportunity to develop a repertoire of approaches to solving complex situations.

Finally, I wish to caution educators against adhering to some of the common linear and cyclical methods that have been heralded for the development of effective problem-solving behaviours. Many of these frameworks are born out of the behaviourist tradition of psychological research (Ohlsson 2012) and fail to take into account what we now know and what continued research in the area is elucidating regarding the effects personal traits and situated and embodied cognitive, sociocultural, epistemological and ontological factors have on the problem-solving process.

References

Arrowsmith-Young, B. (2012). *The woman who changed her brain*. London: Square Peg.

Byrnes, J. P., & Miller, D. C. (2007). The relative importance of predictors of math and science achievement: An opportunity-propensity analysis. *Contemporary Educational Psychology, 32*(4), 599–629.

Delahunty, T. (2014) *Investigating conceptualisation and the approach taken to solving convergent problems: Implications for instructional task design*. Unpublished thesis (Ph.D), University of Limerick.

Delahunty, T., Seery, N., & Lynch, R. (2018). Exploring the use of electroencephalography to gather objective evidence of cognitive processing during problem solving. *Journal of Science Education and Technology, 27*, 114–130.

Delorme, A., & Makeig, S. (2004). EEGLAB: An open source toolbox for analysis of single-trial EEG dynamics including independent component analysis. *Journal of Neuroscience, 134*, 9–21.

Entwistle, N. J. (2000, November). *Promoting deep learning through teaching and assessment: Conceptual frameworks and educational contexts*. Teaching and learning research paper conference, Leicster, England.

Feilzer, M. Y. (2010). Doing mixed methods research pragmatically: Implications for the rediscovery of pragmatism as a research paradigm. *Journal of Mixed Methods Research, 4*(1), 6–16.

Hahn, U., & Chater, N. (1998). Similarity and rules: Distinct exhaustive empirically distinguishable? *Cognition, 65*(2), 197–230.

Jakel, F., & Schreiber, C. (2013). Introspection in problem solving. *Journal of Problem Solving, 6*(1), 20–33.

Jonassen, D. H. (2011). *Learning to solve problems: A handbook for designing problem-solving learning environments*. New York: Routledge.

Jonassen, D. H., Strobel, J., & Lee, C. B. (2006). Everyday problem solving in engineering: Lessons for engineering educators. *Journal of Engineering Education, 95*(2), 139–151.

Kalyuga, S., & Singh, A.-M. (2016). Rethinking the boundaries of cognitive load theory in complex learning. *Educational Psychological Review, 28*, 831–852.

Kitchner, K. (1983). Cognition, metacognition, and epistemic cognition. *Human Development, 26*(4), 222–232.

Markauskaite, L., & Goodyear, P. (2017). *Epistemic fluency and professional education*. Dordrecht: Springer.

McCormick, R., & Davidson, M. (2009). Problem solving and the tyranny of product outcomes. *Journal of Design and Technology Education, 1*(3), 230–241.

O'Donoghue, J., & Kooij, H. V.-D. (2007). *Assessing adults' quantitative skills – The INULIS project* (pp. 251–263). The 14th international conference of adult learning mathematics (ALM), Limerick.

Ohlsson, S. (2012). The problems with problem solving: Reflections on the rise, current status, and possible future of a cognitive research paradigm. *The Journal of Problem Solving, 5*(1), 101–128.

Payne, T. C., Gallagher, K., Eck, J. E., & Frank, J. (2013). Problem framing in problem solving: A case study. *Policing: An International Journal of Police Strategies and Management, 36*(4), 670–682.

Sweller, J. (1988). Cognitive load during problem solving: Effects on learning. *Cognitive Science, 12*, 257–285.

Sweller, J., Ayres, P., & Kalyuga, S. (2011). *Cognitive load theory: Explorations in the learning sciences, instructional systems and performance technologies*. New York: Springer.

Chapter 11
The Influence of Task Difficulty on Engagement, Performance and Self-Efficacy

Jason Power

Abstract My research examined the impact of a person's belief about their own capabilities and how this influences their performance. In order to examine this, I needed a task that was both relatively enjoyable, so that participants would engage with it in their own free time without pressure to do so, and a task that was not heavily linked to a particular subject as this would influence performance. That is the line of thinking that led to a PhD examining self-efficacy theory by getting hundreds of children to play Pacman, a popular arcade game. The research showed that task properties, such as difficulty, can have a considerable impact on how students perceive their own ability. This in turn impacts their likelihood to persist in the face of adversity and ultimately their performance.

The Questions I Asked and Why I Think They Are Important

As my study grew, so too did the questions that I sought to answer. The earlier questions were rather simplistic in nature, but as my understanding of human behaviour and the factors that influence it grew, so too did the nature of the questions.

First I asked how task difficulty impacts an individual's performance. This may seem to have an obvious answer, but often the answers we take for granted are partially, if not entirely, false. As discussed later, difficulty plays a large part in how we see ourselves relative to the activity in question. And how we see ourselves influences everything we do. The delicate balance of an appropriate difficulty level is something that educators are well aware of. Too difficult and the student will disengage almost immediately, while on the other hand, a task that is too easy is often dismissed as not pushing the student's knowledge in the area forward. But if we take the factor of developing knowledge out of the equation, we can get a clearer look at how difficulty is impacting the individual's behaviour. Can a small decrease

J. Power (✉)
University of Limerick, Limerick, Ireland
e-mail: Jason.Power@ul.ie

P J. Williams, D. Barlex (eds.), *Explorations in Technology Education Research*, Contemporary Issues in Technology Education,
https://doi.org/10.1007/978-981-13-3010-0_11

in difficulty result in a big increase in engagement? Can this increase in engagement make up for practising with a task that is objectively easier than the final 'exam'?

This led to the question of how reward influences a person's behaviour. If difficulty remains the same for all, could reward alter behaviour? If increased reward could result in increased engagement, would it not be a superior way to coax engagement in the desired direction? This question also provided additional perspectives on the difficulty question. Ultimately when an individual engages with an easy task, they tend to receive greater rewards, be it a higher percentage in a low-level math test or praise from a parent. So if low difficulty results in increased reward, how can we say what is due to difficulty and what is due to reward? By examining each in isolation, we can begin to form an estimation of each.

These questions in turn led to me asking whether I could cancel out the negative effects of increased difficulty with the positive effects of increased reward. Could I get participants to demonstrate increased engagement with the highest difficulty level? This also provided another snapshot of how difficulty and reward were interacting with each other. But ultimately the questions so far were not getting to the heart of the matter. I was looking at outcomes, scores, engagement rates, etc. But it was clear that it was the way participants viewed their own ability that was the common link between difficulty and reward. This is where I adopted the self-efficacy theory into the study. Self-efficacy is a person's belief in their ability to succeed or accomplish a task in specific situations (Bandura 1986, 1997). A person's sense of self-efficacy can play a major role in how one approaches goals, tasks and challenges. This leads to the final and most important question of the entire study:

How does difficulty influence self-efficacy?

How I Tried to Answer the Questions

I could say that I earned a PhD in Pacman or rather by observing how individuals behaved when playing Pacman. This became the defining description of my research for a few years. As a new postgraduate student, you tend to go to great lengths to impress the very important professors around you, and it is a common practice to attach overly complicated names or descriptions to simple concepts. It was in this manner that I referred to Pacman as a 'digital maze navigation task' for my first 2 years. But I will explain why I believe Pacman, a popular arcade video game (see Fig. 11.1.), was an ideal platform for examining the role of difficulty, reward and self-efficacy when looking at behaviour.

When I set out to examine the influence of difficulty, I was originally going to use math tasks. However there were multiple problems with that idea. People have wide-ranging levels of ability when it comes to math. They have a considerable amount of baggage, both good and bad, and most children that I know would not engage in extra math of their own free will. So I needed a task that did not draw heavily from previously learned skills, that did not come with considerable previous experience and that was reasonably motivating. On top of this, the task needed to be

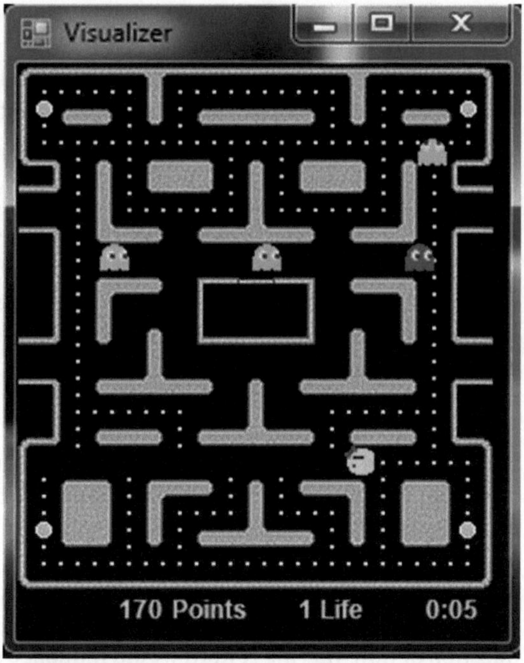

Fig. 11.1 Computerised maze navigation task

able to facilitate various subtle manipulations for the purpose of an experiment. Pacman satisfied all of these requirements and also promised to greatly aid participant recruitment. By slightly altering the game code, different properties could be altered to increase or decrease difficulty or reward.

Over the course of four rounds, participants played Pacman during a school week on their personal USB version of the game. This USB recorded date, time and score whenever it was used. At the start of the week, each participant would play a standard version of the game (to gauge initial ability level) and again at the end of the week (to gauge improvement). What the participants did not know is that they had been given one of three versions of the practice task with a subtle manipulation of either difficulty or reward based on the round. The final round added a Sources of Self-Efficacy scale to examine how these hidden variations in the game were affecting their beliefs surrounding their capability for the task. The completed study involved over 240 students who were all in their first year of second-level education (mean age, 12.4 years).

What I Found Out

The early rounds of the study addressing the impact of difficulty on overall performance proved to be very interesting. One might assume that by practising all week on the highest difficulty level that when the participant sat the final 'exam' using a comparatively easier task, they would perform better than their peers. This however proved not to be the case. Those who practised with the highest difficulty consistently performed worse than the group that practised using either the easy version of the task. Next came the medium-difficulty group and finally, with the highest score, the group that practised with the low-difficulty version of the Pacman practice task.

The impact of difficulty, that very obvious question I asked earlier, proved not to be as straight forward as I had expected. In order to explain the difference in scores, I examined the data stored on the USB sticks. When the amount of practice attempts was calculated per difficulty group, a clear pattern emerged (see Fig. 11.2). Those who practised with the highest difficulty level practised the least amount of times in the first day, and this pattern continued throughout the practice period. Any benefit they might have gained by practising with a task that was more challenging than the final 'exam' was completely overshadowed by the benefit of practice. With those who practised with the easy version of the task, the opposite was true. Any disadvantage this group may have suffered by practicing with a task that was less challenging that the final 'exam' was completely eclipsed by the benefit of increased engagement with the practice task.

This suggested that difficulty was impacting something deeper, altering something that went beyond simplistic models of behaviour. This leads to the question of how varying reward would influence performance. This time participants practised with a high-reward, medium-reward and low-reward version of the task. The reward was the amount of points they received for each power pill they collected in the game. Once again participants were unaware of these hidden variations.

This resulted in patterns that you would expect. The high-reward group showed greater engagement and overall performance, while the low-reward demonstrated the lowest rates of engagement and performance. However the differences between

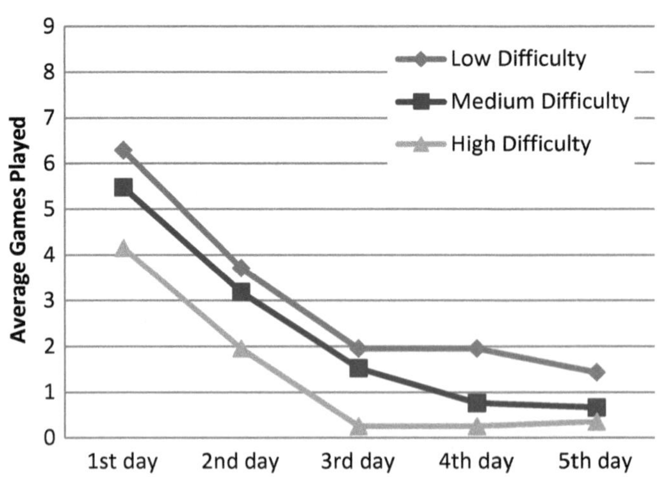

Fig. 11.2 Manipulated difficulty engagement rates

groups were far smaller than those observed in the difficulty variation round suggesting that reward was not as significant a factor as difficulty. In order to further examine the influence of difficulty and reward, a further round was designed. In this round the task was manipulated to see to what degree the positive effects of increased reward could offset the negative effects of increased difficulty. Rather than one negating the other a different pattern emerged.

The expected pattern was the high difficulty being dragged up by the high reward and the low difficulty being dragged down by the low reward. This however was not the case. All groups showed much greater improvements than those observed in previous rounds. While overall the low-difficulty group still improved the most, and engaged with the practice task the most, the high-difficulty/high-reward group scored higher than any group in the previous round. All groups practising with this proportional reward version of the task scored far higher than their peers in previous rounds. This suggested that difficulty and reward were interacting with each other. At this point in the study, it was clear that something far more complex was happening than individuals simply reacting to variations in the task. In terms of psychology, the study to this point could be said to be operating under a behaviourist approach. Behaviourism attributes changes in behaviour almost entirely to environmental factors. It is also concerned almost exclusively with observable events. If the study was stopped after the first two rounds, then we could satisfy the earlier questions by examining them in this manner. Difficulty was altered (environmental factor), this impacted engagement (observable event), and this impacted performance (observable event). However this failed to explain multiple points that had arisen throughout the study. Why had lower difficulty resulted in increased engagement? Why had reward and difficulty interacted to produce such a large increase? These questions could not be satisfied within behaviourism, thankfully a much clever man than I had encountered a similar problem over 40 years before.

When Albert Bandura developed self-efficacy theory (Bandura 1986, 1997), it was in response to what he saw as the shortcomings of behaviourism. He believed that humans were much more complex and that something else was happening when we altered our behaviour. Where behaviourism was only concerned with the interaction between behaviour and the environment, Bandura thought that an additional factor was at play, namely, cognition (Fig. 11.3).

Bandura's earlier work in social learning theory examined how cognition influenced behaviour; he called this triadic reciprocality. His additional factor in the model stressed that there were many internal factors (cognitive skills or attitudes) that influenced behaviour. But just as cognition could influence behaviour (i.e. I believe I can so I do), so too can behaviour influence cognition (i.e. I did before so I believe I can). Environment could also impact behaviour, as we saw in the previous round, but could also influence cognition. It is within cognition that we can picture a person's self-efficacy, their belief in their ability. When considering the influence of self-efficacy (cognition) on performance in Pacman (behaviour), we could yet again class it as having an obvious answer. But if you have been paying attention so far, you may have spotted that all of our obvious answers have proven to be false.

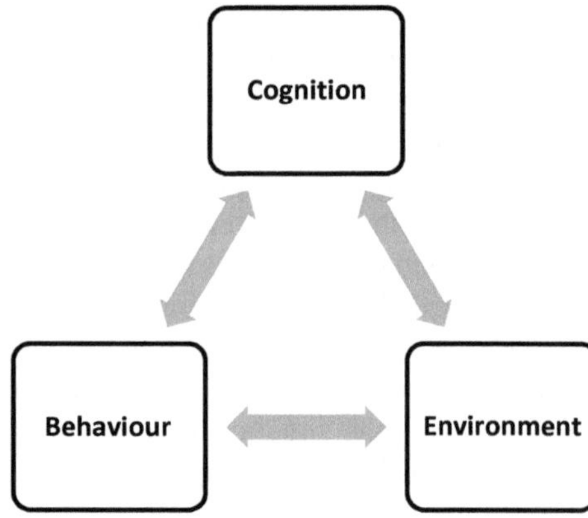

Fig. 11.3 Bandura's triadic reciprocality model

Much of the current research examining self-efficacy and performance has stopped at the obvious. They have looked at how high levels of self-efficacy predict high performance but that only tells part of the story. Does simply having an increased sense of your own ability lead to success? So why were individuals with increased self-efficacy performing better? Could self-efficacy be manipulated in order to increase performance? In order to gain a better understanding of the influence of self-efficacy, a final round was designed to answer the question: *What is the relationship between difficulty and self-efficacy?*

The model in Fig. 11.4 shows how self-efficacy could interact with cognition and fit into the triadic model we looked at previously. Also shown are the four sources of self-efficacy as theorised by Bandura. Individuals draw from these informational sources in order to create their sense of self-efficacy for a given task. Mastery refers to previous positive outcomes in a similar task. Social influence refers to information drawn from people around the individual, for example, positive or negative feedback from a peer, a coach, a parent, etc. Physiological refers to information drawn from your own physical state, for example, rapid heartbeat or sweaty hands before a presentation. Vicarious refers to observing an individual who you feel is close to your level of ability completing a similar task successfully. For example, Mary solved the math problem, and I am about as good at maths as she is. In order to examine this interaction, a final round was conducted.

In order to measure self-efficacy, participants completed a Sources of Self-Efficacy scale after their first game, and just before their last game, a selection of example items from the scale are included in Fig. 11.5.

In this round they practised with the hidden difficulty variations only. Previous patterns of engagement and performance were repeated with those practising with the low-difficulty level scoring the lowest and those practising with the high difficulty levels scoring the highest. However the data from the sources of self-efficacy scale presented a potential explanation for the bigger question, why? The lower difficulty lead to increased engagement, which has been observed in previous rounds. But it also led to higher levels of mastery experience (claimed to be the most

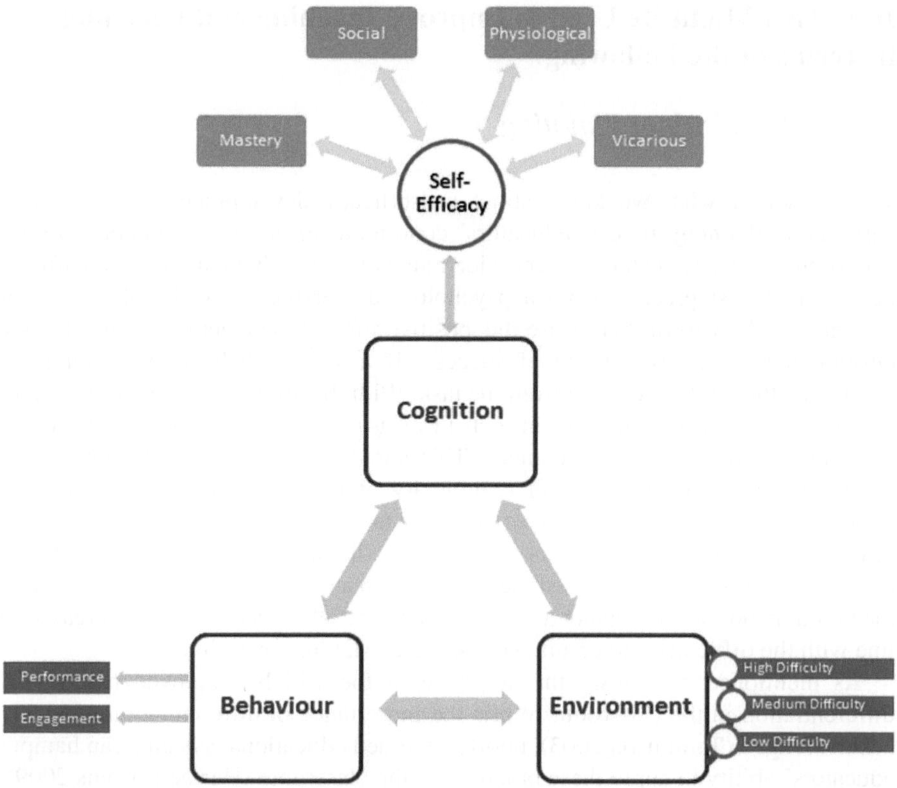

Fig. 11.4 Integration of triadic reciprocality, self-efficacy and task difficulty

6. I do well on even the higher levels of Pacman

Strongly Disagree								Strongly Agree	
1	2	3	4	5	6	7	8	9	10

7. Seeing others do well in Pacman pushes me to do better

Strongly Disagree								Strongly Agree	
1	2	3	4	5	6	7	8	9	10

8. When I see how others win a game, I can picture myself winning the game in the same way

Strongly Disagree								Strongly Agree	
1	2	3	4	5	6	7	8	9	10

Fig. 11.5 Example of self-efficacy scale items

powerful source of self-efficacy), vicarious experience and physiological state. No meaningful change was observed in the physiological source; this source has proven difficult to examine reliably in many studies. Ultimately this final round led to some important conclusions. Self-efficacy could be manipulated by difficulty. It aligned with positive outcomes such as increased engagement and perhaps most importantly highlights the impact of task design on how a child views their own capability.

How This Might Be Used to Improve Teaching and Learning in Terms of the Following:

Overall Curriculum Planning

The impact of what we know about self-efficacy development, in relation to curriculum planning, reflects educators' continuous efforts to differentiate within classrooms that often cater to a very wide range of ability. In many ways it reaffirms pedagogical best practice from a psychological perspective while also offering insights into how to further utilise this positive force. Fundamental to self-efficacy development is the possibility of success. If a student believes that failure is inevitable, they will disengage from the task. With this in mind, it is vital that more complex concepts are scaffolded in such a way as to provide opportunity for success early in the student's first encounter. This should not be confused with simply lowering the overall difficulty or complexity of the curriculum. Rather it is an opportunity for the student to experience mastery, the most potent source of self-efficacy, at an early stage. This reduces the likelihood of disengagement and provides an opportunity for the student to develop a sense of competency relative to the task at hand. In this manner a student's sense of self-efficacy can be increased in line with the difficulty and complexity of the curriculum over time.

As mentioned previously this aligns with the widely employed practice of differentiation in the classroom. While the importance of differentiation is widely acknowledged (Tomlinson 2003), poorly designed educational systems can hamper educators' ability to apply these practices in the classroom (Hertberg-Davis 2009). This highlights two key issues that must be addressed. First educators should be involved in the design of new educational structures such as curriculum or must at a minimum be afforded the professional autonomy to select their own differentiation techniques in a classroom. Secondly it is important that designers of these educational systems are aware of the potential debilitating effects that early and repeated failure can have on a student's self-efficacy. This is especially true for students who are most at risk of dropout and underperformance (Caprara et al. 2008). In addition curriculum designers should be aware of the influences of social persuasions in the development of self-efficacy.

Social persuasions, in the form of feedback from peers, parents and teachers, will influence an individual's self-efficacy, and curriculum designers can take considerations that will facilitate the engagement of this source. Inclusion of project-based learning provides a useful medium for exploiting this source. The conclusion of a project, or a part thereof, provides an opportunity for critical feedback from a variety of sources. In order to amplify this effect, a designer can facilitate structured feedback systems so that peers provide meaningful feedback as opposed to simply praising the individual. As previously discussed, praise should not be confused with critical feedback. Critical feedback will focus on successful outcomes and will tend to be domain specific; this is of much greater value in terms of self-efficacy development. Conversely praise will tend to focus on an attribute of the individual that may have little to do with the domain in question.

Similar considerations can be taken if blended or online learning environments are to be used. Evidence suggests that increased dropout rates and poor engagement rates in online and blended learning environments are linked to self-efficacy (Bates and Khasawneh 2007; Lee 2015; Wang and Newlin 2002). Researchers have identified many different ways in which self-efficacy can influence behaviour in these learning environments. One impediment is low self-efficacy for using the online medium in question. This specifically relates to students' beliefs around their ability to effectively use the online learning systems. This highlights selecting systems that are age appropriate and user friendly. If advanced systems must be used, then adequate training and support must be provided if students are to reach the learning goals that drove the selection and inclusion of the medium. In addition to this barrier, the problem of reduced social interactions in online learning environments has the potential to limit self-efficacy formation for learning. Self-efficacy for learning specifically relates to an individual's belief in their ability to learn the current topic. As you might imagine, this is domain specific. For example, my self-efficacy related to my ability to learn linear algebra is vastly different from my self-efficacy for learning French. Lower social interactions can reduce availability of relevant information from the social persuasion and vicarious experience sources. This can also hamper mastery source input as students often judge success in a task by comparing performance to peers if clear success criteria are not provided.

However the typical classroom use of online platforms more closely resembles a blended learning approach. This is advantageous for a number of reasons. Educators can assess students' capability in terms of using the online system in person and provide additional guidance if necessary. Educators can also facilitate feedback, and peer interaction, in the classroom negating the previous drawbacks of purely online learning environments. You may have noticed that self-efficacy development seems to focus heavily on social interactions. This reflects its origin from Bandura's social cognitive theory. Ultimately judgements surrounding our own competence are entirely dependent on context and comparisons. It is important for curriculum designers to facilitate practitioners in employing strategies that can enhance student development and learning outcomes.

Assessment for Learning

As mentioned previously, an individual develops their sense of self-efficacy by drawing on four informational sources (mastery, social persuasions, physiological and vicarious). The manner in which an individual is assessed will have a significant impact on these sources. Mastery in particular cannot be simplified to a high grade or a level that the teacher may deem successful. Rather it will depend on the individual's perception of their performance. For example, a student who typically receives D grades could deem a B to be a successful outcome; in this scenario the B grade becomes a useful source of mastery, whereas a student accustomed to A grades would most likely deem a B to be an unsuccessful outcome.

However more descriptive feedback surrounding how the individual has improved relative to previous performance is a far better source as it combines mastery with the social persuasion source. As an authority figure and expert, descriptive feedback from you as an educator can have a profoundly positive impact on a student's self-efficacy. However it is important to distinguish meaningful feedback from simple positive compliments. Feedback that focuses on elements of the performance and behaviour that resulted in the positive outcomes, rather than the positive outcome itself, ensures that the student attributes the outcome to effort and behaviour which they have control over. This is in contrast to a belief that the positive outcome is simply related to some natural aptitude for the task. Yeager and Dweck (2012) provide an overview of what they refer to as growth mindset, which is relevant when discussing the impacts of self-beliefs in the classroom. In addition it is worth noting John Hattie's influential meta-analysis studies have repeatedly highlighted the considerable impact of feedback on learning outcomes (Hattie 2013; Hattie and Gan 2011), with some of his more recent work acknowledging the many different types of feedback and how these can influence the learning process (Hattie 2015).

Teaching Methods in Lessons

As noted in previous sections, the decisions that you as an educator make in terms of planning and teaching will have profound impacts on how your students perceive themselves. This in turn will affect behaviour and ultimately learning. Being aware of this impact will allow you to maximise the desired outcomes while simultaneously minimising undesirable outcomes. When designing activities for the classroom, an educator should be cognisant of the previously identified factors which will impact student's self-perceptions, namely, structuring difficulty so that students can experience some form of mastery in the early stages but also in controlling how that student comes to judge mastery experiences.

This requires complex topics to be broken down into smaller manageable activities while employing differentiation to increase difficulty on an individual basis. By preparing multiple difficulty levels in advance, teachers will allow themselves time to focus in class time to address students who may be struggling. Once a student with low self-efficacy has been identified, you employ the social persuasion source as a means of bolstering self-efficacy through targeted feedback. In addition you can employ the vicarious source by pairing a student who is struggling with the current level with another student. This will allow for peer teaching but also provides a chance for the student in question to observe a peer succeeding. It is important that the student considers this peer roughly comparable in order for the vicarious source to be effective. If the student simply believes the other student is naturally excellent, then the vicarious source will provide little benefit. Ultimately educators need to be aware of the pitfalls of low self-efficacy more so than the benefits of exceptionally high self-efficacy. Students who believe they have no chance of succeeding in the given task will have poor motivation and can often be a continuous source of behavioural issues. Providing early opportunities for success is more about the impact on future behaviour than the initial comparatively shallow learning outcome.

Other Areas Related to Your Research

Self-efficacy is only one of many self-regulation theories. These theories examine how individuals behave and think relative to certain tasks and activities and have been studied in areas as diverse as professional sports, business and lifestyle. The reason I believe teachers should consider these theories when examining their own practice is that research shows that we have a huge impact on how our students see themselves. This is a responsibility that we should not take lightly. It has the potential to increase motivation, self-esteem and multiple desirable behaviours. Conversely, if we act with little regard to the impacts of our professional decisions and actions, we risk increasing dropout, disengagement and increasing behavioural issues in our classrooms. It is vital that we as educators appreciate the influence that we have on our students and that the lessons we teach go far beyond the prescribed topics of a subject. Teaching at its most important level is a profession which aims to develop people. By understanding how people behave relative to your actions, you can help to develop behaviours and cognition that have the potential to help an individual in a wide range of areas that go beyond your chosen subject discipline. With that in mind, I have found the following texts to be informative, accessible and of practical use for the modern classroom. Artino (2012) provides an excellent overview of self-efficacy theory and its relevance to classroom practice. Although the paper was written with medical educators in mind, I believe it is one of the best pragmatic discussions of the benefits and potential misunderstandings of the theory that has been written to date. For a broader discussion examining the influence of mindsets on students' behaviour, I would recommend Yeager and Dweck (2012). Dweck's work has been hugely influential, and her work surrounding growth mindset is compatible and complimentary to self-efficacy theory.

What Else Would Be Good to Know, and How Could Teachers Find Out?

Self-efficacy theory belongs to a larger group of theories collectively referred to as self-regulation theories. These theories focus on how an individual exerts control over their own behaviour. For us, as educators, it is hard to overestimate the importance of self-regulation in our classrooms. It impacts how and what we teach, but most importantly it can have profound impacts on the development of the individual child. For example, whether a person chooses to persevere or give up after failing at a task may seem like an insignificant act in isolation but over time can have devastating impacts on a child's educational experience. This is why it is important to examine theories such as self-efficacy, not as abstract works created for their own sake but as limited glimpses into an endlessly complex system that influences human behaviour. What I find personally interesting about the research discussed here is the conclusion that we as educators, through the decisions we make and practices we engage in, can alter students' views of their own capabilities. By extension we can influence how they see themselves as a person. Ultimately I believe this is the most important responsibility of an educator, the personal development of our students. This has many implications for classroom practice. If, as the research discussed here suggests, students' beliefs surrounding their ability have such impacts on their learning, why is it that we almost exclusively focus assessment almost entirely on learning outcomes? By extension why not directly target these areas through interventions in our classrooms? An ongoing research project, PERTS (Project for Education Research that Scales) [Access at www.perts.net], provides accessible research and materials that educators may use in their classroom in order to assess students' beliefs. Perhaps more importantly they outline interventions that have been rigorously designed and are supported by a considerable body of evidence and which have been designed for use in a classroom.

In the previous description of research and its relevance to your role as an educator, I focused heavily on self-efficacy; however this is only one theory among many. There are numerous competing self-regulation theories, and ultimately it is up to the practitioner to evaluate each (Sitzmann and Ely 2011). While theories within self-regulation may be described as competing, there are many other educational and psychological theories that could be described as complimentary to self-efficacy. I have previously mentioned Dweck's mindset theory which I believe is both relevant and compatible, but Dweck has also created a very useful overview of self-theories and their role in motivation, personality and development (Dweck 2000). If I were to conclude with one message that summarises the previous discussion surrounding research, theory and practice, it would be this. You as an educator influence how your students perceive themselves. This can have considerable impacts on student behaviour and development, both negative and positive.

References

Artino, A. R. (2012). Academic self-efficacy: From educational theory to instructional practice. *Perspectives on Medical Education, 1*(2), 76–85.

Bandura, A. (1986). *Social foundations of thought and action: A social cognitive theory.* Englewood Cliffs: Prentice Hall.

Bandura, A. (1997). *Self efficacy: The exercise of control*. New York: Freeman.

Bates, R., & Khasawneh, S. (2007). Self-efficacy and college students' perceptions and use of online learning systems. *Computers in Human Behavior, 23*(1), 175–191.

Caprara, G. V., Fida, R., Vecchione, M., Del Bove, G., Vecchio, G. M., Barbaranelli, C., & Bandura, A. (2008). Longitudinal analysis of the role of perceived self-efficacy for self-regulated learning in academic continuance and achievement. *Journal of Educational Psychology, 100*(3), 525–537.

Dweck, C. S. (2000). *Self-theories: Their role in motivation, personality, and development*. Philadelphia: Psychology Press.

Hattie, J. (2013). *Visible learning: A synthesis of over 800 meta-analyses relating to achievement*. London: Routledge.

Hattie, J. (2015). The applicability of visible learning to higher education. *Scholarship of Teaching and Learning in Psychology, 1*(1), 79.

Hattie, J., & Gan, M. (2011). Instruction based on feedback. In E. Mayer & P. Alexander (Eds.), *Handbook of research on learning and instruction* (pp. 249–271). New York: Educational Press.

Hertberg-Davis, H. (2009). Myth 7: Differentiation in the regular classroom is equivalent to gifted programs and is sufficient: Classroom teachers have the time, the skill, and the will to differentiate adequately. *Gifted Child Quarterly, 53*(4), 251–253.

Lee, C. Y. (2015). Changes in self-efficacy and task value in online learning. *Distance Education, 36*(1), 59–79.

Sitzmann, T., & Ely, K. (2011). A meta-analysis of self-regulated learning in work-related training and educational attainment: What we know and where we need to go. *Psychological Bulletin, 137*(3), 421.

Tomlinson, C. A. (2003). *Fulfilling the promise of the differentiated classroom: Strategies and tools for responsive teaching*. Alexandria: Association for Supervision and Curriculum Development.

Wang, A. Y., & Newlin, M. H. (2002). Predictors of web-student performance: The role of self-efficacy and reasons for taking an on-line class. *Computers in Human Behavior, 18*(2), 151–163.

Yeager, D. S., & Dweck, C. S. (2012). Mindsets that promote resilience: When students believe that personal characteristics can be developed. *Educational Psychologist, 47*(4), 302–314.

Part IV
Cognition

Chapter 12
Design Cognition and Student Performance

Greg J. Strimel

Abstract The teaching of engineering in US schools has seen a surge in popularity since the turn of the twenty-first century, and as design is considered a defining characteristic of engineers, the practice of engineering design has become a critical component of technology education. Consequently, research related to design cognition in engineering/technology education has become more prevalent in the literature. However, there are often minimal discussions on bridging design research with practice. Therefore, this chapter will present a design cognition research methodology developed to help inform engineering/technology education practice, the results of a study employing this method, and the implications for teaching and learning.

The Questions I Asked and Why They Are Important

Engineering/technology education (ETE) has a history of providing students with opportunities for developing and applying higher-order thinking skills and conceptual knowledge through the resolution of multifaceted problems. Since the launch of the *Standards for Technological Literacy* (ITEA/ITEEA 2000/2002/2007) and the *Next Generation Science Standards* (NGSS Lead States 2013), the prominence of the problem-solving practices of engineering design in the United States has continued to increase, and engineering design activities have become a core feature of both ETE (Asunda and Hill 2007) and science education. While there seems to be broad support for incorporating engineering design activities within various STEM programs (National Academy of Engineering [NAE] and National Research Council [NRC] 2014; Strimel et al. 2016), the learning of such multidimensional concepts and development of the non-routine problem-solving abilities of engineering design are not thoroughly understood (NAE and NRC 2009).

G. J. Strimel (✉)
Purdue University, Lafayette, IN, USA
e-mail: gstrimel@purdue.edu

P J. Williams, D. Barlex (eds.), *Explorations in Technology Education Research*, Contemporary Issues in Technology Education,
https://doi.org/10.1007/978-981-13-3010-0_12

While advocates posit that engineering design activities encourage learners to make new and useful connections among disciplinary-specific knowledge through self-directed learning (NAE and NRC 2014), research suggests that such open-ended learning experiences may pose challenges to students and educators (Berland 2013). For example, some investigations imply that design experiences can distract students from identifying and learning the difficult-to-understand discipline-specific concepts (Goldstone and Sakamoto 2003; Kaminski et al. 2009). As such, a heavy reliance on engineering design activities or the improper scaffolding of such experiences may impede a student's ability to transfer knowledge to other situations. Berland and Busch (2012) state that students can sometimes resolve design problems without an in-depth conceptual understanding of the relevant concepts involved. Thus, Antony (1996) believes that teachers may be falsely comforted by providing students numerous design activities with an expectation that students are successfully constructing transferable knowledge from these experiences. Nevertheless, the increased understanding of student thinking during engineering design tasks in relationship to their design performance can help to pinpoint concerns and enact interventions to enhance a student's learning through design activities (Bransford and Vye 1989; Strimel 2014a).

As the teaching of engineering design continues to expand (Strimel and Grubbs 2016), it becomes critical to understand the ways in which students mentally process engineering design tasks in an effort to provide effective instruction, establish the adequate scaffolding engineering design experiences, and enact interventions to enhance student design abilities (Grubbs 2016). However, there has been minimal research to provide insight into the range of cognitive processes employed by students while designing and making a product, system, or device (Kelley 2008) and limited discussion on effective ways to link design cognition research with practice in ETE (Grubbs and Strimel 2016). As a result, this study was driven by the overarching questions of (a) how do secondary students cognitively navigate an engineering design task from design conception through prototype production and (b) how can we inform ETE by examining design cognition in respect of student performance. Consequently, the following research objectives were established:

RO_1: Identify the fundamental cognitive processes secondary engineering students use to design, make, and evaluate physical prototypes to engineering design problems.

RO_2: Determine potential identifiers within secondary engineering student cognitive processes for monitoring aptitude to successfully design, make, and evaluate physical prototypes.

RO_3: Create a conceptual engineering problem-solving model integrating student cognitive processes for the improved development of design abilities.

How I Tried to Achieve the Objectives

To achieve the stated objectives, eight purposefully selected participants enrolled in advanced high school engineering coursework were tasked to verbalize their thoughts while completing a design task. These participants were required to complete the activity independently to better capture their individual thought processes. To facilitate data collection, participants were equipped with point-of-view cameras that gathered videos of their nonverbal cues and verbalized thought processes. These videos and the participants' design artifacts (design journals, prototypes, and teacher-scored project rubrics) enabled triangulation of data while conducting data analyses (Cross 2004). The engineering design task, provided in Fig. 12.1, was developed to facilitate the collection of the participants' thought processes while they designed, made, and evaluated a physical prototype or prototypes of a solution. In doing so, data were collected as participants progressed through a full design process spectrum. Capturing the cognitive processes involved in the full spectrum of designing enabled the researcher to compare each participant's cognitive strategies with the effectiveness of their final solution. This method can be viewed as an approach toward bridging the gap between design cognition research and teaching practice as the results can help determine cognitive difficulties that hinder a student from developing successful solutions and support the identification of design heuristics for enhancing design capabilities. Consequently, the design task was developed in such a way that also enabled the collection of quantitative data on the effectiveness of each participant's solution. In this situation, effectiveness was measured by the reduction of the turbidity of a contaminated water sample. These quantitative data were then used to help determine potential relationships between solution effectiveness and the mental strategies employed by participants.

Following data collection, the video recordings of each participant's design process (average of 1 h and 50 min in length) were divided into three distinct phases of designing, making, and evaluating. These videos were then segmented into individual cognitive tasks and coded using the 17 mental processes for technological problem-solving defined and validated by Halfin (1973). The segmenting and coding process enabled the researcher to determine how long and how often each participant employed each mental process. The operational definitions of these processes are provided in Table 12.1. However, based on a review of literature, the mental process of *Modeling* was determined by the researcher to be too similar to the other codes of *Model/Prototype Constructing* and *Designing*. As a result, the *Modeling* mental process was not used as a code, and the actions that could be considered *Modeling* were coded as either *Designing* or *Model/Prototype Constructing*.

Once each video was coded, the frequencies and the amount of time each participant employed each coded cognitive process were calculated. To ensure the reliability of this procedure, two coders independently coded each participant's protocols, and a Pearson's r correlation coefficient between the coding results of the two coders was calculated. These results revealed a high level of agreement between the two coders.

ENGINEERING DESIGN CHALLENGE

Problem Statement

People in developing countries do not have continuous access to clean water, especially after the onset of a natural disaster. Water in these situations needs significant purification. However, water purification units are expensive and not easy to obtain. Therefore, you are tasked to design an inexpensive, easy to use, easy to assemble, durable, and low maintenance water purification system using low cost, readily available materials to quickly remove contaminants from water. You will focus on reducing the turbidity of a sample of water.

Testing Performance

Turbidity is a measure of the lack of clarity (cloudiness) of water and is a key test of water quality. Turbidity is apparent when light reflects off of particles in the water. Sources of turbidity include soil erosion, waste discharge, urban runoff, events that stir up sediments, and other organic compounds that result from decay of leaves and plants, and algal growth. In addition to creating an unappealing cloudiness in drinking water, turbidity can be a health concern. It can sustain or promote the regrowth of pathogens in the water distribution system, which can lead to the spread of waterborne diseases. Turbidity is measured in Nephelometric Turbidity Units, NTU. Water is visibly turbid at levels above 5 NTU. The standard for drinking water is below 1.0 NTU.

Materials	Equipment	Deliverables
You can use any materials necessary to create the best solution. (Typical filter materials selected by students included: charcoal, gravel, sand, filter paper, fabric, burlap, and cotton)	• Computer and Internet Access • Distilled Water • Contaminated Water • Sample Bottle • Vernier Turbidity Sensor/Equipment and Software	• Functioning Prototype of "Quality" Construction • Project Journal • Solution Justification - A summary of the details of the design, its benefits, uses, and other important information that explains the design solution.

Sample Student Solution #1

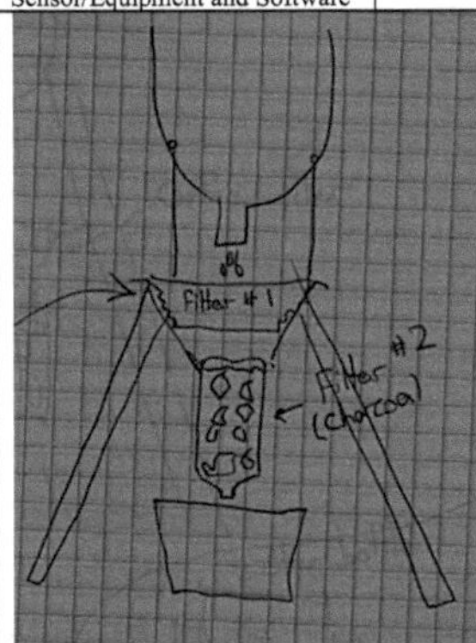

Sample Student Solution #2

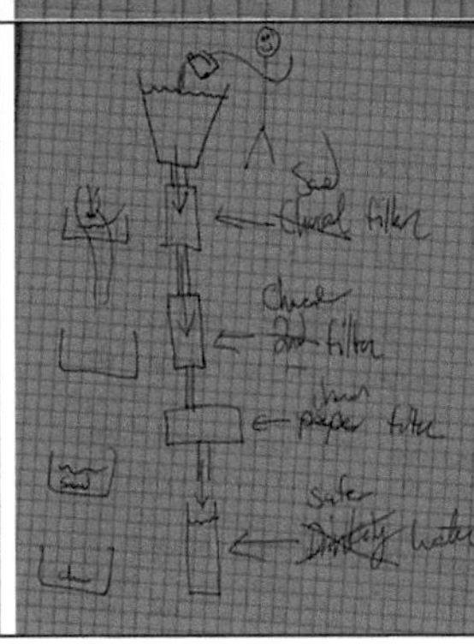

Fig. 12.1 Engineering design challenge

What I Found Out

The first research objective was to identify the fundamental cognitive processes advanced secondary engineering students use to design, make, and evaluate physical solutions to an engineering design task. This objective was met by coding videos of participants thinking aloud during an engineering design session using Halfin's (1973)

Table 12.1 17 original mental processes for solving technological problems

Cognitive process	Definition
Analyzing (AN)	This is the process of identifying, isolating, taking apart, breaking down, or performing similar actions for the purpose of setting forth or clarifying the basic components of a phenomenon, problem, opportunity, object, system, or point of view
Communicating (CM)	This is the process of conveying information (or ideas) from one source (sender) to another (receiver) through a media using various modes (the modes may be oral or written or pictures or symbols or any combination of these)
Computing (CP)	This is the process of selecting and applying mathematical symbols, operations, and processes to describe, estimate, calculate, quantify, relate, and/or evaluate in the real or abstract numerical sense
Creating (CR)	This is the process of combining the basic components or ideas of phenomena, objects, events, systems, or points of view in a unique manner that will better satisfy a need, either for the individual or for the outside world
Defining problem(s) (DP)	This is the process of stating or defining a problem, which will then enhance the investigation leading to an optimal solution. It is transforming one state of affairs to another desired state
Designing (DE)	This is the process of conceiving, creating, inventing, contriving, sketching, or planning by which some practical ends may be affected, or proposing a goal to meet the societal needs, desires, problems, or opportunities and do things better. Design is a cyclic or iterative process of continuous refinement or improvement
Experimenting (EX)	This is the process of determining the effects of something previously untried in order to test the validity of an hypothesis, to demonstrate a known (or unknown) truth, or to try out various factors relating to a particular phenomenon, problem, opportunity element, object, event, system, or point of view
Interpreting data (ID)	This is the process of clarifying, evaluating, explaining, and translating to provide (or communicate) the meaning of particular data
Measuring (ME)	This is the process of describing characteristics (by the use of numbers) of a phenomenon, problem, opportunity, element, object, event, system, or point of view in terms that are transferable. Measurements are made by direct or indirect means, are on relative or absolute scales, and are continuous or discontinuous
Modeling (MO)	This is the process of producing or reducing an act or condition to a generalized construct that may then be presented graphically in the form of a sketch, diagram, or equation; physically in the form of a scale model or prototype; or in the form of a written generalization
Model/prototype Constructing (MP)	This is the process of forming, making, building, fabricating, creating, or combining parts to produce a scale model or prototype
Observing (OB)	This is the process of interacting with the environment through one or more of the senses (seeing, hearing, touching, smelling, or tasting). The senses are used to determine the characteristics of a phenomenon, problem, opportunity, element, object, event, system, or point of view. The observer's experiences, values, and associations may influence the results

(continued)

Table 12.1 (continued)

Cognitive process	Definition
Predicting (PR)	This is the process of prophesying or foretelling something in advance, anticipating the future based on special knowledge
Questioning/ hypothesizing (QH)	Questioning is the process of asking, interrogating, challenging, or seeking answers related to a phenomenon, problem, opportunity, element, object, event, system, or point of view
Testing (TE)	This is the process of determining the workability of a model, component, system, product, or point of view in a real or simulated environment to obtain information for clarifying or modifying design specifications
Visualizing (VI)	This is the process of perceiving a phenomenon, problem, opportunity, element, object, event, or system in the form of a mental image based on the experience of the perceiver. It includes an exercise of all the senses in establishing a valid mental analogy for the phenomena involved in a problem or opportunity

Note. Halfin (1973), Hill and Wicklein (1999), and Wicklein and Rojewski (1999)

17 mental processes for technological problem-solving. On average, the participants completed the design task within 1 h and 51 min. Throughout the entire engineering design activity, the top three most employed mental processes were *Model/Prototype Constructing*, *Analyzing*, and *Managing*. *Model/Prototype Constructing* consumed 23.3% of the participant's time on average, which mostly consisted of physically manipulating materials. Next, *Analyzing* consumed 15.8% of the participant's time on average and consisted mostly of conducting research and analyzing the effectiveness of their design decisions. Lastly, *Managing* consumed 13.9% of their time on average and consisted mostly of the participants planning their actions and gathering necessary resources. The least used mental processes were *Experimenting* (0.7%), *Computing* (0.08%), *Questioning/Hypothesizing* (1.9%), *Defining Problems* (2.0%), *Interpreting Data* (2.1%), *Predicting* (2.3%), and *Measuring* (2.6%). The limited use of these mental processes may be a reflection of the curricula and instructional strategies used in ETE programs. The overall average of each mental process employed by the participants can be found in Table 12.2.

When examining the participant data by the phases of designing, making, and evaluating, participants dedicated the majority of their time to the evaluation phase ($\bar{x} = 41$ min and 21.1 sec), followed by making ($\bar{x} = 39$ min and 45.7 sec) and designing ($\bar{x} = 29$ min and 29 sec). These results indicate that participants dedicated a limited amount of time to planning their designs before physically making their prototypes. The lack of planning was observed as the main reason why participants required more time for the evaluation phase. The omission of thorough design practices required more efforts to fix and refine their prototype. Therefore, the participant's approach resembled a trial-and-error method of problem-solving over an informed and intentional engineering design tactic.

However, during the designing phase, *Analyzing*, which consisted of dissecting information to apply it toward developing a design concept, was employed the most—consuming 41.1% of the participant's time on average. Additionally,

Table 12.2 Mean total cognitive processes used throughout the engineering design activity

Entire engineering design session		
Cognitive process	$\bar{x}$ times used	$\bar{x}$ amount of time
Analyzing	49.625	17:27.5
Communicating	33.125	06:30.5
Computing	6.125	00:52.5
Creating	18.250	02:54.9
Designing	27.750	08:00.8
Defining problems	7.125	02:12.6
Experimenting	3.125	00:49.3
Interpreting data	18.500	02:21.8
Managing	56.500	15:24.8
Measuring	16.000	02:51.1
Modeling	1.125	00:09.8
Model/prototype constructing	47.875	25:49.8
Observing	31.375	04:48.5
Predicting	30.250	02:30.2
Questioning/hypothesizing	16.000	02:05.9
Testing	28.500	12:20.0
Visualizing	25.625	03:32.5
Total	**416.875**	**1:50:35.8**

Note. $\bar{x}$ represents the sample mean

Designing was used an average of 18.2% during the participant's designing phase. *Model/Prototype Constructing* was used the most during the making phase, as the participants took an average of 54.3% of their time physically building their solutions, and *Testing* was used the most during the solution evaluation phase. See Figs. 12.2, 12.3, and 12.4 for an illustration of the average use of each mental process during each phase of the design process.

Research Objective 2 was to determine potential identifiers within student cognitive processes for monitoring aptitude to successfully design, make, and evaluate physical solution prototypes. To achieve this objective, the participants' solution effectiveness was tested and the resulting data were used to rank which solution solved the problem most successfully. Next, the data from the top- and bottom-performing participants were used to determine the mean percentage of time taken for each mental process by both groups during the engineering design session. Overall, the top-performing participants took more time solving the engineering design challenge than the bottom-performing participants and held a higher total frequency of employing the various mental processes. The top participants took 15 min and 59.9 sec longer to solve the problem and employed 226 more total frequencies of the mental processes.

Throughout the engineering design session, the top participants devoted the majority of their problem-solving time to employing the cognitive process of *Analyzing* (18.6%) information, *Testing* (17.0%) their solutions, *Managing* (14.3%) their problem-solving process, and *Communicating* (9.4%) their designs/results.

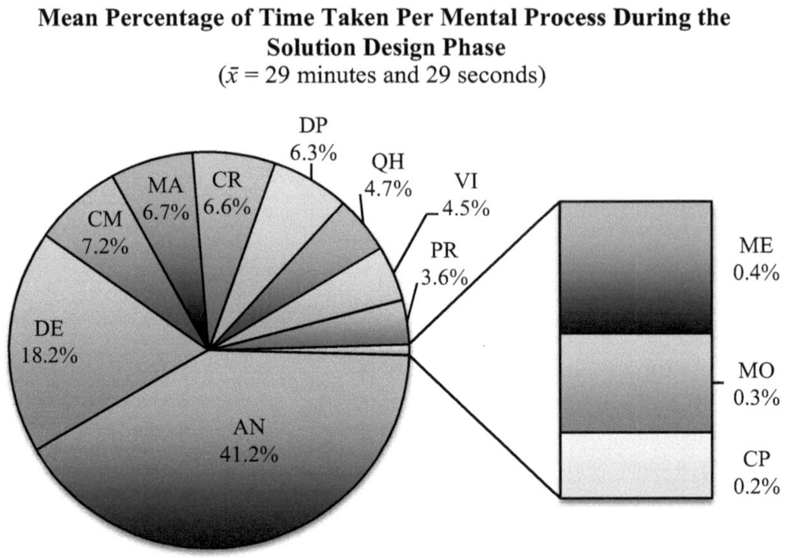

Fig. 12.2 Mental process used during the solution design phase

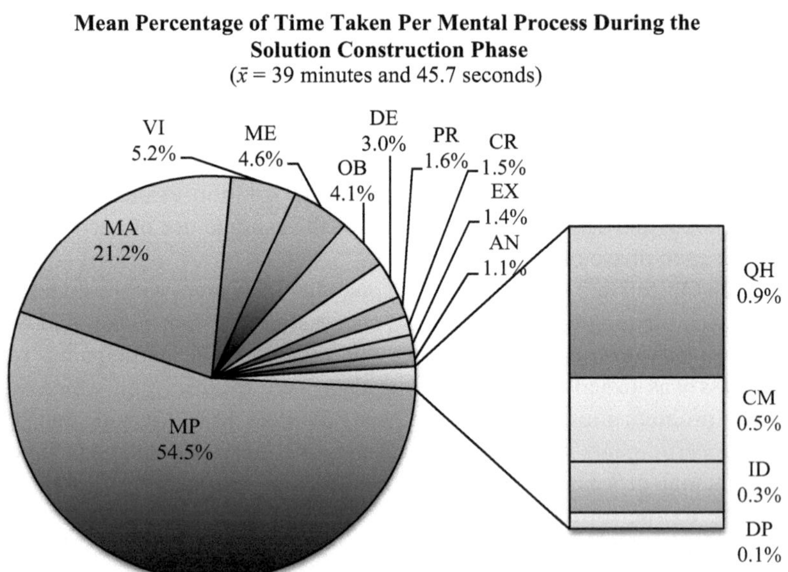

Fig. 12.3 Mental process use during the solution-making phase

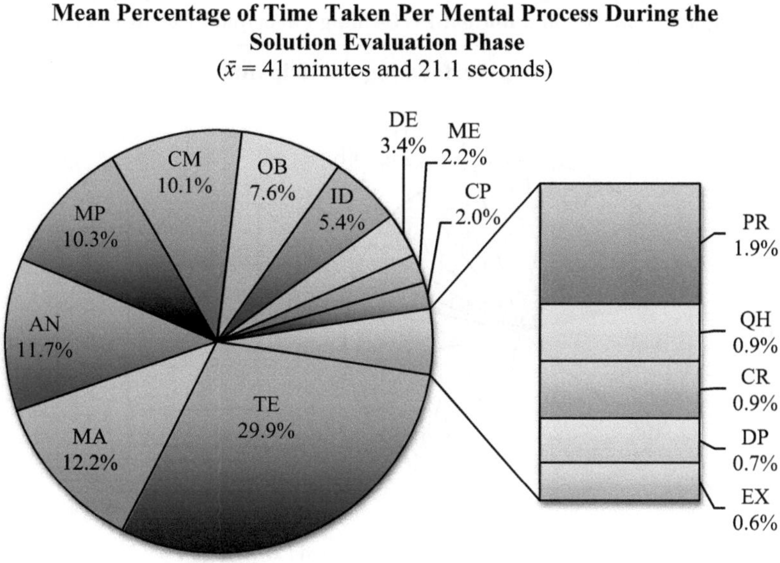

Fig. 12.4 Mental process use during the solution evaluation phase

However, the bottom participants took the majority of their problem-solving time employing the processes of *Model/Prototype Constructing* (29.8%), *Analyzing* (15.9%), *Managing* (12.7%), and *Designing* (8.4%). One major difference between these two groups were the percentages of time taken employing the processes of *Testing* and *Model/Prototype Constructing*. The top participants took 17.0% of their time *Testing* and re-*Testing* their solutions, while the bottom participants only took 5.9% of their time employing this mental process. This difference is a reflection of how the bottom-performing participants were not observed iteratively testing, improving, and re-testing their solutions. Additionally, the bottom participants were observed focusing more of their time on constructing their prototypes, while the top participants were observed only dedicating 8.9% of their time for employing the process of *Model/Prototype Constructing*. A complete graphical comparison of the average percentage of time taken for each process used by the top- and bottom-performing participants during the entire engineering design session is reported in Fig. 12.5.

The data for the top- and bottom-performing participants were also analyzed and compared at the three different phases (design, make, and evaluate) of the engineering design process. The data analysis indicated that the top-performing participants took 7 min and 31.2 sec more than the bottom-performing participants during the designing phase. During this phase, the top participants devoted 9.2% more of their time *Communicating* than the bottom participants. The top participants also took 8.1% more of their solution design time employing the *Managing* process than the bottom-performing participants. However, the bottom-performing participants dedicated 8.1% more of their solution design time to *Defining the Problem* and 3.4%

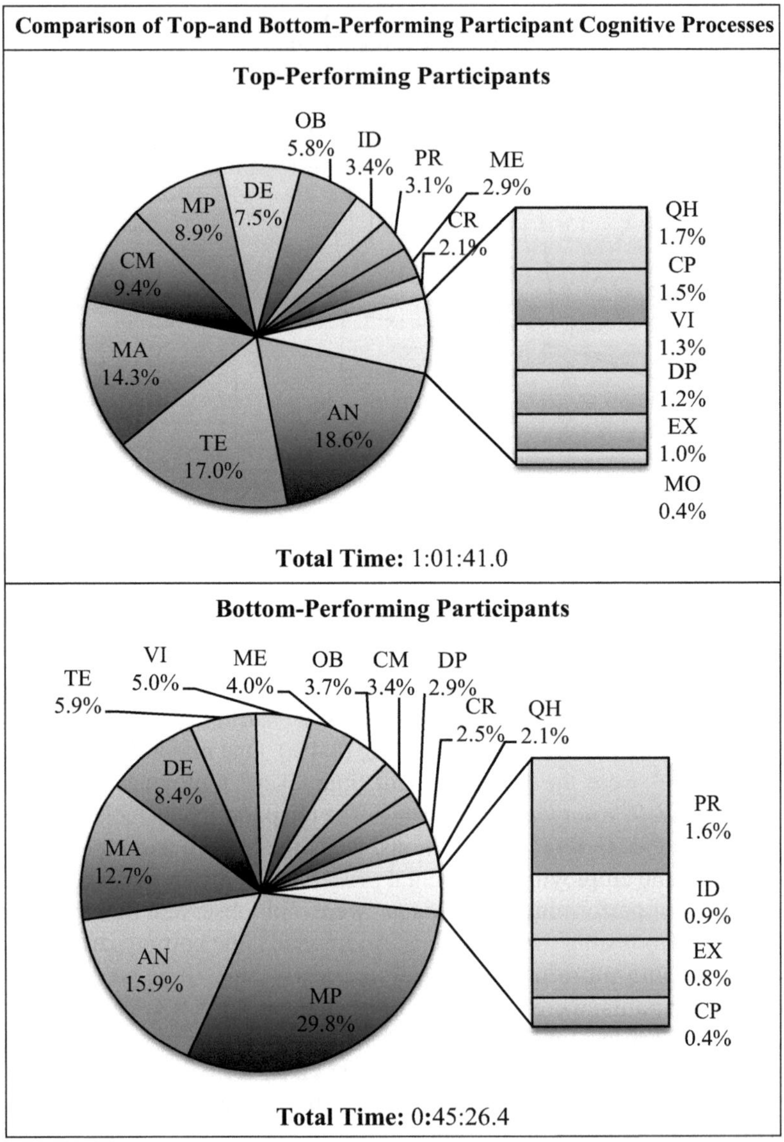

Fig. 12.5 The comparison of top- and bottom-performing participant cognitive processes

more time *Questioning/Hypothesizing*. Furthermore, the teacher rubric evaluations of the participants' work indicated that the top-performing participants scored the same as the bottom-performing participants on the category of *Research*. However, the top-performing participants scored 55% higher in the category of *Multiple Solutions* and 35% higher in the *Design Justification* category.

During the making phase, the bottom-performing participants took 38 min and 15.6 sec longer building their prototype than the top-performing participants. However, the percentage of time taken per mental process was similar between both groups. Furthermore, teacher rubric evaluations of the participants work indicated that the top- and bottom-performing participants received very similar scores during this phase. The top participants scored a total of 10% higher than the bottom participants on the three categories of *Material Selection*, *Prototype Durability*, and *Prototype Use* combined.

During the evaluation phase, the top-performing participants took 46 min and 44.2 sec more time assessing and improving their solutions than the bottom-performing participants. The top participants expended a greater amount of time using the *Model/Prototype Constructing* process during this phase to iteratively refine their prototypes than the bottom-performing participants. Additionally, the teacher rubric evaluations of the participants' work indicated that the top participants scored 50% higher in the *Prototype Performance* category, 15% higher in the *Prototype Testing* category, 50% higher in the *Prototype Revision* category, and 25% higher in the *Engineering Documenting* category than the bottom-performing participants.

Research Objective 3 was to create a conceptual engineering problem-solving model integrating student cognitive processes for the improved understanding and development of problem-solving abilities. This objective was achieved by examining the participant observations and the corresponding coded data to create graphical representations of the procedure that each participant employed to solve the engineering design task. These representations were also checked against participant-generated concept maps of their design procedure. The analysis of the mapped cognitive process employed in each case study led the researcher to identify two distinctly different approaches to solving the engineering design problem. Some participants followed a methodical, sequential process for solving the problem, while other participants followed a more unformulated, nonsequential trial and error process.

The sequential participants planned and followed a very logical step-by-step process for creating a solution to the problem and conducted multiple iterations of testing, redesigning, and re-testing until they reached a desired outcome. These participants were more focused on the problem definition and meeting the established solution criteria versus physically building a prototype. Additionally, these participants were proactive when it came to addressing issues that arose when creating their solutions. Conversely, the nonsequential participants followed a less structured, trial-and-error approach to develop a solution without following a particular plan. These participants were more reactive in terms of confronting issues when creating their solutions. The nonsequential participants were focused more on the physical building of a prototype rather than developing a complete design plan for creating and evaluating their solution. Finally, the researcher used the graphical representations of these two design approaches combined with the participant cognition data related to solution effectiveness to generate a conceptual engineering design model. This model is provided in Strimel (2014a) on pages 153 through 160.

How This Might Be Used to Improve Teaching and Learning

According to Barak and Hacker (2011), the enhancement of ETE curriculum and instruction is contingent upon the role researchers and educators take in developing and implementing an understanding of the cognitive processing of students when creating effective solutions to engineering design problems. In addition, the mental processes employed by students, or lack thereof, can be used as indicators of voids in curricula, instruction, and student learning. Therefore, the results of this study can provide opportunities for improving the way in which ETE curriculum and instruction are developed (Kelley 2008) as well as informing the assessment of a students' design competencies. Based upon the findings highlighted, the researcher developed the following recommendations for the teaching and learning of authentic engineering design practices.

Exercise Project and Process Managing Strategies

The mental process of *Managing* was one of the most used by participants throughout their engineering design sessions. The majority of the participants' time dedicated to *Managing* consisted of directing their actions during the design session and controlling the inputs for solution development. However, some participants were more effective at planning than others. Novice designers tend to lack the self-discipline to develop a comprehensive plan for project completion and rely more on a trial-and-error approach to solving a problem. Whereas, expert designers may be more likely to dedicate time toward developing a well-thought out problem-solving strategy. Nonetheless, the findings indicated that effective planning and project management is a possible contributor toward creating a more successful engineering solution. Therefore, ETE curricula should include effective project management and process planning techniques when implementing engineering design-based lessons. Some potential design heuristics to be practiced can include separating the project into manageable tasks; establishing benchmarks for project completion; evaluating the available resources; developing a timeline using project management tools (i.e., a Gantt chart); creating a team charter to establish team direction, responsibilities, and boundaries; and formulating a detailed work plan before progressing in their engineering design process. According to Portz (2014), practicing the planning and managing of design projects can provide students with critical skills for authentic future workplace settings.

Increase Focus on Computational Thinking and Modeling

Computational thinking can be referred to as the combination of disciplinary knowledge and the necessary thought processes to formulate problems in a manner that enables one to use computers and other information-processing agents to effectively solve them (International Society for Technology in Education and Computer Science Teachers Association 2011). Computational thinking skills and practices have now been embraced by STEM educators at all levels of schooling (Magana and Coutinho 2017). These skills and practices are key features of engineering design and enable a designer to use quantitative data to estimate, calculate, or describe how well a design concept solves a particular problem. However, in the study described in this chapter, the mental processes (*Computing, Measuring,* and *Interpreting Data*) that closely align with computational thinking and support the development of computational models were the least employed by the participants during the engineering design sessions. This can be problematic as these processes are often considered fundamental and authentic engineering-related practices for predicting whether certain design decisions will be successful or not. According to Grubbs and Strimel (2015), it is these computational practices that can separate engineering design from general trial-and-error problem-solving. Therefore, ETE should emphasize the application of developmentally appropriate mathematical practices, computational modeling techniques, and quantitative data analyses during engineering design challenges.

The *Committee for the Workshops on Computational Thinking* (NRC 2011) posits that computational thinking practices are vital to integrate within engineering education as engineers habitually rely on computational models to analyze designs and scientific principles constructed through visualizations of data. As a result, ETE students should be exposed to explicit integrated examples of age-appropriate mathematics and computational tools (i.e., Microsoft Excel, MATLAB, CAD) while interacting with an engineering design challenge. To ensure this recommendation is achieved, curriculum should demonstrate how computational thinking skills can support the optimization of engineering solutions and require the application of computer programming logic and principles as part of the problem-solving process. Providing ETE students support for developing logic and writing code to perform computations, analyze data, plot data, and create simulations of solutions may help deepen high school students' engineering design experiences, enhance their design optimization abilities, and ultimately, aid them in developing better performing design solutions.

Attention to Quality (Proper Making Abilities)

Opportunities to study and comprehend the proper selection and use of tools, materials, and equipment/software were historically core features of ETE that have vanished from many primary and secondary programs (Cool et al. 2017). Due to a variety of factors, many ETE programs have transitioned to using low-cost, low-tech tools/materials as the main sources for making physical prototypes (Grubbs 2014). Potentially as a consequence, the observations in this study indicated an absence of students selecting the proper tools and equipment when developing a solution, a lack of focus on constructing quality prototypes, a minimal amount of time employing the mental process of *Measuring*, and the improper use of tools, equipment, and materials. The participants seemed to pay little attention to precisely planning their designs, adding dimensions to their solution sketches, and accurately measuring the materials they used to make their prototypes. As a result, participants wasted materials and time by only making estimations when manipulating materials and relying on mending materials, such as hot glue and duct tape, to correct any production errors they encountered. However, participants with enhanced tool knowledge and capabilities did tend to have better performing prototypes. Therefore, as stated by Cool et al. (2017), it is vital for ETE to "provide experiences to better practice "making" as well as performing informed design decisions based on the properties of materials and their abilities to manipulate them" (p. 16).

It is recommended that curriculum developers highlight the importance of carefully producing quality prototypes and generate engineering design challenges within a laboratory setting that scaffolds the appropriate tool knowledge and technique (Cool et al. 2017; Haynie 2008, 2009; Strimel et al. 2016). To ensure this recommendation is achieved, students should practice (a) evaluating and selecting industry quality tools/equipment when developing a solution, (b) developing a bill of materials for the proposed solution based on material properties, (c) using software and advanced measurement tools to generate detailed technical visual representations of designs, (d) establishing and comprehending design specifications, and (e) determining authentic testing procedures that enables the collection of valuable data to inform design revisions. In doing so, ETE teachers can provide more authentic learning experiences for students and provide them with the skills for producing viable physical solutions to engineering problems. As Lombardi (2007) claims, authentic learning experiences should culminate in the creation of a *polished* product, valuable in its own right, and students should know what it feels like to be held accountable for these products.

Emphasize Scientific Inquiry and Experimentation to Inform Design

As Orr and Flowers (2014) claim, emphasizing experimentation in the ETE classroom can enable students to learn through inquiry and to use experiment results to refine their design concepts. Accordingly, the findings of this study highlighted that participants who devoted more time employing the scientific-related mental processes (*Testing*, *Experimenting*, *Observing*, and *Interpreting Data*) produced better performing solutions. The participants were observed using these mental processes to iteratively refine their designs by setting up small-scale experiments to test different ideas or materials, collect data not readily available for their specific context, and interpret the data in a manner that informed their design modifications. Therefore, it is recommended that ETE curricula integrate the use of more scientific inquiry methods within engineering design challenges. This can be valuable for students to develop experiments to gather pertinent information to inform their design or test and evaluate design concepts.

Orr and Flowers (2014) recommend that teachers include experimentation in their curriculum and instruction as much as possible through teacher-directed experiments, student-selected experiments, student-found topics, and student inquiry because it promotes students to use evidence to inform their problem-solving process. One method to support these scientific practices during design is to implement a science education instructional strategy known as the POE (Predict-Observe-Explain) method. Ayvaci (2013) describes the POE method as an instructional technique for students to predict the results of a scientific investigation based on sound reasoning, conduct an experiment or trial while making observations, and develop explanations of the results to eliminate the incongruities between predictions and observations. This can be a powerful tool for the purpose of informing design decisions through scientific inquiry. For example, when designing a device to reduce the turbidity of contaminated water, students may pose a driving question related to their design such as "what is the best order for arranging the sequence of filter materials?" Following the POE method, students would then develop a prediction and provide a rationale by connecting with their prior knowledge. Next, students would determine what information they need to gather to answer their question and establish the way in which to collect the necessary data. Lastly, students would document their observations and develop an explanation of the results to support or reject their prediction in order to inform their design decisions. When using this approach while designing, one can optimize design concepts (a key engineering design practice) prior to making a physical prototype—potentially saving time and resources.

Provide Time for Iteration

The top-performing participants in this study enacted more iterations of testing their solutions, making observations, interpreting the outcome data, and then experimenting with design changes to improve their solutions. The bottom-performing participants were focused on making the solution and were satisfied after testing their solutions only once. Therefore, it is recommended that instructors model the appropriate behaviors of iteratively testing solutions, properly analyzing data, and utilizing the resulting data to make design optimizations. Additionally, instructors should allocate time for students to reflect upon their results. Strimel (2014b) believes that reflecting on prototype testing can extend a students' learning and enhance their problem-solving abilities. However, time for iteration can often be omitted from ETE curricula. As a result, many learning opportunities may be missed by not reinforcing the authentic engineering practices for iteration in ETE curricula.

Alignment with Engineering Practices

The data highlighted some potential disconnects between ETE curricula and the engineering-related professions. The actions and thoughts of the participants did not regularly coincide with authentic engineering practices. The data indicated that the participants were focused on physically building their solutions and took a relatively minimal amount of time thoroughly planning their designs before beginning the production of their solution. Little time was used for analytical design and modeling, and many of the participants did not utilize testing data to optimize their designs. Additionally, most of the participants did not experiment with materials to determine what would be the best choice for their solution; instead they relied on repair materials. This may indicate that the engineering habits of mind, which involve design, analysis, modeling, and optimization, are not stressed or accurately practiced throughout ETE curriculum and instruction. Therefore, it is recommended that ETE programs clarify their aim and establish their program's purpose. If their purpose is to teach engineering principles, then it should be important to align with the best practices of the engineering design.

Balance Context and Content

The participants who created the best-performing solutions noted in their design journals and reflections that they related the challenge to some other experiences they have had. Whether it was experiences using certain materials/tools or experiences with a similar platform for making solutions, it seemed to help them direct their problem-solving process. This information is similar to the results of an analysis of practicing engineers conducted by Jonassen et al. (2006). Their findings indicated that drawing upon prior experiences is the most important factor in solving an engineering workplace problem. Based on this finding, it is recommended that ETE curricula provide students with specific authentic tools, materials, and design experiences that can be drawn upon to solve other problems in the future. In doing so, ETE can distinguish itself from other school subjects by providing authentic situational contexts in which students can have experiences with realistic materials, advanced prototyping technologies, and appropriate resources to solve the most relevant engineering challenges. Moving away from generic problem-solving skills to focusing on engineering design practices in relevant contexts can potentially support the transfer of learning to future problem contexts. However, this recommendation highlights the question of valuing the context of learning over specific content and requires one to think about balancing the trade-offs between rich contexts and amounts of content to be learned.

What Else Would Be Good to Know and How Could Teachers Find Out

Design is often touted as the core practice of engineers. Moreover, engineering design, a specific type of design, is considered the central feature of engineering and thus now a core feature of ETE. However, when teaching engineering design, it is impractical to think students will be able to comprehend and enact authentic engineering design practices without the proper scaffolding of design experiences. Therefore, one must question how to properly scaffold design activities to provide a coherent transition to, and study of, authentic and rigorous engineering design practices across the grade levels. Yet, primary, secondary, and post-secondary teachers often deliver similar design activities across the grades. For instance, students in the middle grades are often tasked with building a bridge to span a distance while holding the most weight. This activity can also be seen in high school curriculum and even in college-level introductory engineering coursework. This can be a result of the lack of coherency in ETE and the absence of scaffolding design experiences along with the developmentally appropriate STEM knowledge. Likewise, engineering design process models are regularly prescribed in classrooms for students to follow in these design experiences. However, these models are often generated based on the opinion of experts and do not take into account the cognitive development of students at various age levels (Strimel 2014a). Consequently, teachers of these curricula may deliver the design process models without the proper introduction to and scaffolding of the design skills/practices. In doing so, students may never experience the suitable application of design, mathematical, and scientific principles to generate multiple solution ideas and narrow down the ideas through the engineering practices of predictive analysis and computational modeling (Harris and Jacobs 1995; Merrill et al. 2009). Therefore, a final recommendation is to use design cognition research results, such as those presented in this chapter, to better inform the scaffolding of design experiences to support a coherent study of engineering principles and design practices in ETE. There have been several design cognition research articles studying primary and secondary student design cognition published that can be valuable to inform the teaching and learning of design. Grubbs et al. (2018) provide an overview of the design cognition literature that has been conducted at the primary and secondary levels of education that can be used to inform teaching and learning.

References

Antony, G. (1996). Active learning in a constructivist framework. *Educational Studies in Mathematics, 31*(4), 349–369.

Asunda, P. A., & Hill, R. B. (2007). Critical features of engineering design in technology education. *Journal of Industrial Teacher Education, 44*(1), 25–48 Retrieved from http://scholar.lib.vt.edu/ejournals/JITE/v44n1/pdf/asunda.pdf.

Ayvaci, H. S. (2013). Investigating the effectiveness of predict observe explain strategy on teaching photo electricity topic. *Journal of Baltic Science Education, 12*(5), 548–564.

Barak, M., & Hacker, M. (2011). Learning theories for engineering and technology education. In M. Barak & M. Hacker (Eds.), *Fostering human development through engineering and technology education* (pp. vii–vxi). Rotterdam: Sense Publishers.

Berland, L. K. (2013). Designing for STEM integration. *Journal of Pre-College Engineering Education Research, 3*(1), 22–31.

Berland, L. K., & Busch, K. (2012). Negotiating STEM epistemic commitments for engineering design challenges. *American Society for Engineering Education. pp 00856-19. 2012.,* 00856-00819.

Bransford, J. D., & Vye, N. J. (1989). A perspective on cognitive research and its implications for instruction. In L. Resnick & L. E. Klopfer (Eds.), *Toward the thinking curriculum: Current cognitive research* (pp. 173–205). Alexandria: ASCD.

Cool, N., Strimel, G. J., Croly, M., & Grubbs, M. E. (2017). Making mason beehives: Teaching proper "making" skills through authentic engineering design contexts. *Technology and Engineering Teacher, 76*(8), 20–25.

Cross, N. (2004). Expertise in design: An overview. *Design Studies, 25*(5), 427–441.

Goldstone, R. L., & Sakamoto, Y. (2003). The transfer of abstract principles governing complex adaptive systems. *Cognitive Psychology, 46*(4), 414–466.

Grubbs, M. E. (2014). Genetically modified organisms: A design-based biotechnology approach. *Technology and Engineering Teacher, 73*(7), 24–29.

Grubbs, M. E. (2016). *Further characterization of high school pre- and non-engineering students' cognitive activity during engineering design.* Doctoral dissertation. Retrieved from ProQuest dissertations and theses database. (UMI No. 3662376).

Grubbs, M. E., & Strimel, G. (2015). Engineering design: The great integrator. *Journal of STEM Teacher Education, 50*(1), 77–90.

Grubbs, M. E. & Strimel, G. J. (2016, June 26). Cognitive research: Transferring theories and findings to k-12 engineering educational practice. *American society for engineering education 103rd annual conference and exposition.* New Orleans, LA.

Grubbs, M. E., Strimel, G. J., & Kim, E. (2018). Examining design cognition coding schemes for P-12 engineering education. *International Journal of Design & Technology Education*, 1–23. https://link-springer-com.ezproxy.lib.purdue.edu/content/pdf/10.1007%2Fs10798-017-9427-y.pdf.

Halfin, H. H. (1973). *Technology: A process approach*. Doctoral dissertation, West Virginia University, 1973. Dissertation abstracts International, (1) 1111A.

Harris, T. A., & Jacobs, H. R. (1995). On effective methods to teach mechanical design. *Journal of Engineering Education, 84*(4), 343–349.

Haynie, W. J. (2008). Are we compromising safety in the preparation of technology education teachers? *Journal of Technology Education, 19*(2), 94–98.

Haynie, W. J., III. (2009). Safety and liability in the new technology laboratory. *Technology Teacher, 69*(3), 31–36.

Hill, R. B., & Wicklein, R. C. (1999). A factor analysis of primary mental processes for technological problem-solving. *Journal of Industrial Teacher Education, 36*(2), 83–100.

International Society for Technology in Education & Computer Science Teachers Association. (2011). *Operational definition of computational thinking for K-12 education*. Retrieved from http://www.iste.org/docs/ct-documents/computational-thinking-operational-definition-flyer.pdf?sfvrsn=2

International Technology and Engineering Educators Association (ITEA/ITEEA). (2000/2002/2007). Standards for technological literacy: Content for the study of technology. Reston: Author.

Jonassen, D. H., Strobel, J., & Lee, C. B. (2006). Everyday problem-solving in engineering: Lessons for engineering educators. *Journal of Engineering Education, 95*(2), 139–151.

Kaminski, J. A., Sloutsky, V. M., & Heckler, A. (2009). Transfer of mathematical knowledge: The portability of generic instantiations. *Child Development Perspectives, 3*(3), 151–155.

Kelley, T. R. (2008). Cognitive processes of students participating in engineering-focused design instruction. *Journal of Technology Education, 19*(2), 50–64.

Lombardi, M. (2007). *Authentic learning for the 21st century: An overview*. Retrieved from http://net.educause.edu/ir/library/pdf/eLi3009.pdf

Magana, A. J., & Coutinho, G. S. (2017). Modeling and simulation practices for a computational thinking-enabled engineering workforce. *Computer Applications in Engineering Education*. https://doi.org/10.1002/cae.21779.

Merrill, C., Custer, R. L., Daugherty, J., Westrick, M., & Zeng, Y. (2009). Delivering core engineering concepts to secondary level students. *Journal of Technology Education, 20*(1), 48–64.

National Academy of Engineering and National Research Council. (2009). *Engineering in K-12 education*. Washington, DC: National Academies Press.

National Academy of Engineering and National Research Council. (2014). *STEM integration in k-12 education: Status, prospects, and an agenda for research*. Washington, DC: The National Academies Press. https://doi.org/10.17226/18612.

National Research Council. (2011). *Report of a workshop on the pedagogical aspects of computational thinking*. Washington, DC: The National Academies Press.

NGSS Lead States. (2013). *Next generation science standards: For states, by states*. Washington, DC: National Academies Press.

Orr, T., & Flowers, J. (2014). An experimental approach to everything. *Technology and Engineering Teacher, 73*(8), 8–12.

Portz, S. (2014). Teaching project management. *Technology and Engineering Teacher, 73*(7), 19–23.

Strimel, G. (2014a). Authentic education by providing a situation for student-selected problem based learning. *Technology and Engineering Teacher, 73*(7), 8–18.

Strimel, G. J. (2014b). *Engineering design: A cognitive process approach*. Doctoral dissertation. Retrieved from ProQuest Dissertations and Theses database. (UMI No. 3662376).

Strimel, G., & Grubbs, M. E. (2016). Positioning technology and engineering education as a key force in STEM education. *Journal of Technology Education, 27*(2), 21–36.

Strimel, G. J., Grubbs, M. E., & Wells, J. G. (2016). Engineering education: A clear decision. *Technology and Engineering Teacher, 76*(4), 18–24.

Wicklein, R. C., & Rojewski, J. W. (1999). Toward a "unified curriculum framework" for technology education. *Journal of Industrial Teacher Education, 36*(4), 38–56.

Chapter 13
Design Cognition: Strategies for Teachers in Practice

Michael E. Grubbs

Abstract Developing students' higher-order thinking capability is a fundamental goal of education. Thus, design cognition has been recommended as a viable approach to address this need. Yet, current research findings primarily illustrate how students think, not pedagogical strategies for enhancing students' design thinking. Thus, the purpose of this chapter is to bridge research and practice, by presenting strategies and heuristics, for instructors, on how to provide feedback through assessments (pre, formative, and summative) and how design challenges can be used to situate students within a context that intentionally and appropriately develops their cognitive abilities.

Questions Posed to Improve Design Cognition Teaching and Learning

Of the many goals of STEM educational reform, one is to expose K-12 students to the T and E of STEM. The intent is to provide a vehicle for establishing connections between subject matter, improve student learning, and enhance their cognition (NAE & NRC 2009). In turn, engineering design is specifically recognized as a viable learning approach that can develop students' cognition and address current goals of STEM educational reform. Therefore, the rationale for the study stems from the need to understand students' cognitive processes during these experiences (Barak and Hacker 2011; Petrina 2010; Ritz and Martin 2012; Strimel 2014; Zuga 2004). Increased understanding of students' cognitive processes during design can allow educators to identify and correct student learning problems (Bransford and Vye 1989; Strimel 2014).

Although researchers have examined the design cognition of professional designers and students at the collegiate level, few research investigations have investigated

M. E. Grubbs (✉)
Baltimore County Public Schools, Towson, MD, USA
e-mail: mgrubbs@bcps.org

P J. Williams, D. Barlex (eds.), *Explorations in Technology Education Research*, Contemporary Issues in Technology Education,
https://doi.org/10.1007/978-981-13-3010-0_13

high school students' cognitive activity during designing (Crismond and Adams 2012; Hynes 2012; Lammi and Becker 2013). Moreover, as researchers have begun to address this gap, they have employed coding schemes comprehensively, using such broad categories as formulation, analysis, and synthesis. However, previous researchers have further characterized these broad descriptors of cognitive activity at a more detailed level (Grubbs 2016; Purcell et al. 1996). Subsequently, what have yet to be understood are K-12 students' cognitive processes in terms of underlying mechanisms during design at a more detailed level.

This chapter extends research that examined high school pre- and nonengineering students' cognitive processes, specifically the underlying mechanisms of broadly used classifications such as problem formulation, analysis, synthesis, and evaluation. Describing students' cognitive processes more clearly established a greater understanding of students' cognitive activity during design, thereby addressing the lack of research-based findings with what educators can do to enhance students' design skills and learning ability (Crismond and Adams 2012). Additionally, these research findings resulted in greater understanding of novice designers' cognitive processes, which can support teachers, in developing instructional tools and assessments that are grounded in findings of cognitive science. In turn, educators can more effectively cultivate students' twenty-first century cognitive competencies and better equip students to deal with challenging problems they will encounter throughout their life (Razzouk and Shute 2012).

The following research questions were used to address the problem and purpose of this study.

1. How do 12th grade high school pre- and nonengineering students distribute their cognitive efforts as they dissect the problem domain during engineering design tasks?
2. How do 12th grade high school pre- and nonengineering students distribute their cognitive efforts as they synthesize, analyze, and evaluate during engineering design tasks?
3. Do differences in cognitive processes emerge between 12th grade high school students who take pre-engineering courses and those who do not?

Solving the Problem

To answer the research questions of the study, an exploratory sequential mixed methods design was chosen. The rationale for choosing this method was to employ a qualitative phase that explored and categorized the cognitive processes of the participants, on a phenomenon that has had little research conducted on it (Creswell 2013). Additionally, the need to quantify the data required the computation of descriptive and inferential statistics to answer research questions 1, 2, and 3.

To begin, this research investigation examined extant data made available through a previously conducted National Science Foundation-funded longitudinal study supported under Grant No. 1160345 that explored and illustrated the cognitive activity of high school junior- and senior-level students who took formal pre-engineering courses and those who did not (NSF 2012; Wells et al. 2014, 2016). Findings reported from the first year of this longitudinal research study, and initial findings of the second year, suggested a lack of cognitive effort during the problem formulation stage. Also, students in both groups were reported as distributing most of their cognitive efforts during the analysis, synthesis, and evaluation phases. Two separate coding schemes, the *problem domain: degree of abstraction* presented in Appendix 1 and the *strategy classification scheme* presented in Appendix 2, are two mechanisms that further illustrated students' cognitive processes during designing and indicated cognitive differences between student groups, not previously identified. The remaining sections describe the research design and procedures for re-examining the original study using these additional coding schemes.

For this study, 40 participants from the second year were selected as a purposeful sample from the entire 2-year set of data from the original NSF research investigation. This sample resulted in 20 dyads consisting of 10 pre-engineering dyads that had taken pre-engineering courses and 10 dyads that had not. The extant data set containing student dyad utterances during the engineering design challenge of year 2 was obtained in transcript form with approval from the previous researchers to further analyze for this study. Participants also gave approval from the initial study to engage in the research project. Data was captured verbatim from the original verbalized thoughts of each student participant. Students were tasked with solving a design-no-make challenge within 45 minutes. They solved the following task and were requested to discuss their thought processes and provide their design process as illustrations on a white board:

> Your design team has been asked to design an assistive technology device that will help a person who is wheelchair bound grab an average size glass food jar off of a cupboard shelf too high for them to reach. The *Reach-n-Grab* device should be designed so that a wheelchair bound person would be able to reach and grab the glass jar when it is about two feet beyond their normal reach.

In addition to the lead researcher, the data of this study was segmented, coded, and arbitrated by three coding analysts. Two of the analysts were graduate research assistants who had experience segmenting, coding, and arbitrating. The third analyst had minimal experience with verbal protocol analysis but had conducted similar qualitative research projects. Prior to the training session, literature related to both coding schemes (Purcell et al. 1996) and how to conduct verbal protocol analysis studies were sent to the analysts to prepare themselves.

Potential Solutions to Understanding Design Cognition

Research Question 1

For research question one, findings illustrated how 12th grade high school pre- and nonengineering students distribute their cognitive efforts as they dissect the problem domain during engineering design tasks. Figure 13.1 provides an illustration of the three codes used to describe students' cognitive processes, within this domain.

As Fig. 13.2 depicts, the first group, *engineering students*, allocated the majority of their efforts, 54.21%, on design issues, related to **details** of the design task, 30.21% of their effort on **subsystems**, and 15.57% on the overall **system** phase. In comparison, as Fig. 13.3 illustrates, the *nonengineering* students similarly devoted the same degree of cognitive effort, in terms of priority to the **detail** phases of design. However, the *nonengineering* students, captured in Fig. 13.3, distributed their cognitive effort fairly evenly between the **system** and **subsystem** phases. Consequently, both student groups exerted more cognitive effort toward the finer details of the design task. This may have included the material properties, specific components (e.g., nuts, bolts, gears), and even colors of individual parts. Also they exerted less cognitive effort toward the actual functionality of the design solution and, overall, the functionality of the system as a whole. As Figs. 13.2 and 13.3 illustrate, the groups are very comparable, yet the engineering group exerted less cognitive effort on the overall system. Meaning, they cognitively considered how the system or details actually worked together and function more than the nonengineering group.

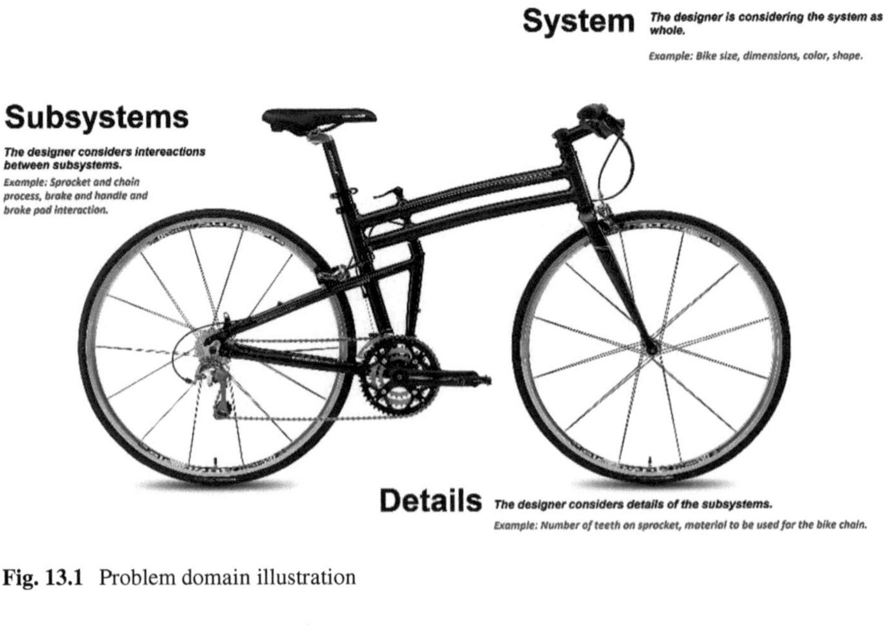

Fig. 13.1 Problem domain illustration

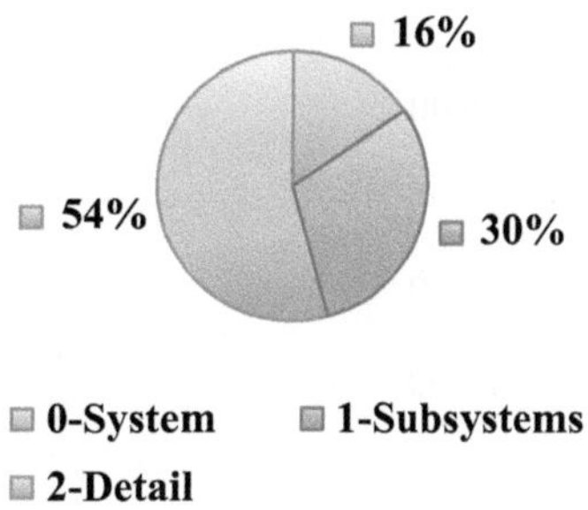

Fig. 13.2 ENG students CS # 1: problem domain – degree of abstraction

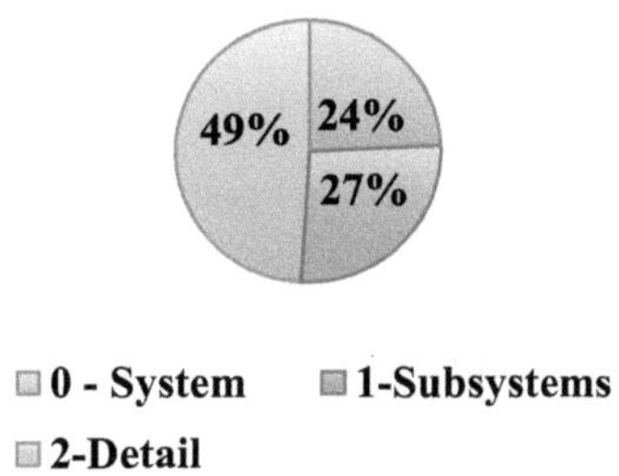

Fig. 13.3 NON students CS # 1: problem domain – degree of abstraction

Research Question 2

The second of the two main research questions that guided this study was how 12th grade high school pre-engineering (ENG) and non-pre-engineering (NON) students distribute their cognitive efforts as they analyze, synthesize, and evaluate during engineering design tasks. To address research question two, the *cognitive strategies* coding scheme was used to interpret students' cognitive issues for each group of participants. Additionally, to address this research question, the qualitative analysis used through coding was the foundation for the quantitative analysis. For example, the computed frequency counts for each of the 19 design issues (Appendix 2) were used to calculate percentage occurrences to characterize students' cognitive efforts while in the cognitive strategies domain. Subsequently, the calculated descriptive statistics are presented in the following section to describe the engineering students, as well as the pre-engineering students' cognitive effort. Figure 13.4 is included to begin to note emerging relationships between the two groups of students, which will be addressed in greater detail for research question three. Essentially, Fig. 13.4 illustrates the 19 issues and compares how each student group exerts effort in relation to those issues.

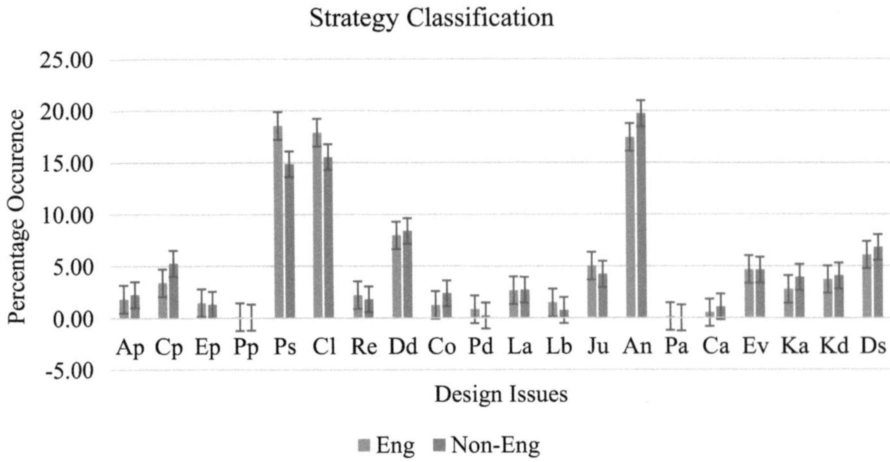

Fig. 13.4 Percent occurrence of strategy classification design issues: ENG vs. NON high school seniors

Pre-engineering Seniors (ENG)

To characterize the cognitive strategies of high school seniors who took pre-engineering courses (ENG), percentage occurrences were calculated from the total number of coded segments for each design session. Verbalized utterances that were not captured within the cognitive strategies coding scheme were removed from total segmentation. Computation of these percentages illustrates how each student group, of the ENG group, distributed their cognitive effort across the various design issues. Once each dyad's cognitive effort was computed, the overall means and standard deviations were calculated for all 19 individual codes. As Fig. 13.4 illustrates, senior year ENG students exerted the most cognitive effort on three design issues: proposing a solution (Ps) at 19%, clarifying a solution (Cl) at 18%, and analyzing a proposed solution (An) at 17%. Collectively, those three design issues cover 54% of their cognitive effort throughout the engineering design task. Conversely, postponing the analysis of the problem (Pp) was one of the lowest cognitive issues, at 0.11%, postponing an analysis action (Pa) at 0.14%, and performing calculations to analyze a proposed solution (Ca) at 0.52%.

There does appear to be one similar trend among dyads of the ENG group that dyads would engage in presenting a proposal for a solution, provide clarification for a solution they have proposed, and analyze a proposed solution. Subsequently, these three design issues are spread across the synthesizing solution and evaluating solution domains, and not the analysis problem and explicit strategies of the cognitive strategies coding scheme.

Non-Pre-engineering Seniors (NON)

The second component of research question two was to characterize how non-pre-engineering (NON) students distribute their cognitive efforts in regard to the coded segments of the cognitive strategies design issues. The percentage occurrences were also calculated from the total number of cognitive strategies coded segments for each dyad. Once each dyad's cognitive effort was computed, the overall means and standard deviations were calculated for all 19 individual codes. As Fig. 13.4 illustrates, senior NON-pre-engineering students exerted the most cognitive effort on analyzing a proposed solution (An) at 20%, clarifying a solution (Cl) at 16%, and proposing a solution (Ps) at 15%. What should be noted is that these three design issues were also the three highest of the ENG group. Conversely, the lowest three cognitive issues that the NON group attended to were postponing an analysis action (Pa) at 0.00%, postponing the analysis of the problem (Pp) at 0.07%, and postponing a design action (Pd) at 0.21%. However, though similar, the ENG group included performing calculations to analyze a proposed solution (Ca) but did not include postponing a design action (Pd) as their lowest. There does appear to be one similar trend among dyads of the NON group that was also evident for the ENG group. That is, dyads attributed a majority of their cognitive effort to presenting a proposal for a solution, clarifying a solution, and analyzing a proposed solution.

Research Question 3

The third and final research question of this study was whether there were differences in cognitive processes between 12th grade high school students who take pre-engineering courses and those who do not. To address this question, a t-test was used to test the difference between the means of the two groups. To compute this statistic, the means and standard deviations were used. This research question has two components that will be reported: the differences between cognitive efforts in the problem domain: degree of abstraction, followed by the differences between strategy classification.

Problem Domain: Degree of Abstraction

As Fig. 13.5 illustrates, and research question one and two alluded to, differences were apparent across all three design issues of the problem domain: degree of abstraction with the most evident difference in the system category (0). Findings indicate that the NON group of seniors spent more cognitive effort considering the problem from the point of view of the user than the ENG group. This additional attention resulted in less cognitive effort in the subsystems and detail design issues than the ENG students. Table 13.1 captures the results of the t-test conducted to determine if the differences illustrated in Fig. 13.5 were statically significant.

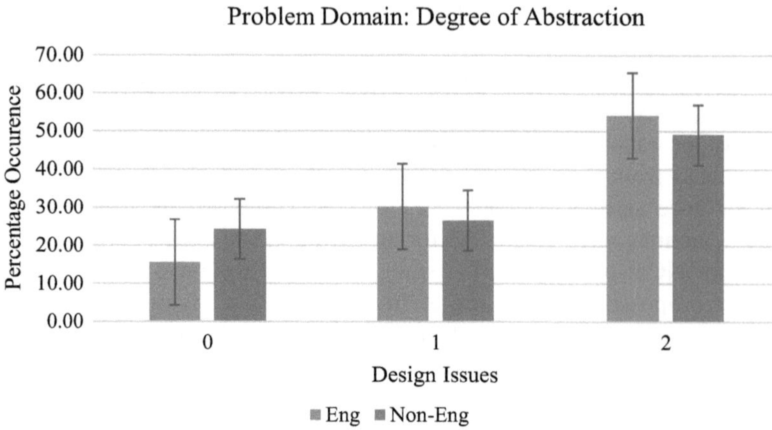

Fig. 13.5 Percent occurrence of problem domain: degree of abstraction design issues: ENG vs. NON high school seniors

Table 13.1 Statistical results of cognitive issues: ENG vs. NON high school seniors

Cognitive issue	t – value (%)	p – value
0-system	−2.67	*0.01*
1-subsystem	0.80	0.22
2-details	−1.06	0.15

Note. Italic value indicates statistically significant difference between ENG and NON groups

Comparisons of the ENG and NON group data using a t-test revealed significant differences which were evident only among how the designer considered the problem from the point of view of the user (System – 0). However, analysis revealed no significant differences for the other two cognitive issues of considering the problem in terms of the subsystems (Subsystems – 1) and considering the details of the subsystems (Details – 2). Therefore, findings from data analysis indicate that the ENG group dyads overall exerted less cognitive effort toward the system issues of the design task than did the nonengineering students.

Strategy Classification

As Fig. 13.4 illustrates, differences were also apparent across multiple design issues of the strategy classification coding scheme. The most obvious design issue differences were proposing a solution (Ps), clarifying a solution (Cl), analyzing a proposed solution (An), consulting information about the problem (Cp), postponing a design action (Pd), and looking back (Lb). Additionally, there appears to be a clear alignment of three strategies (e.g., Ps, Cl, and An) between the 15 and 20 percentage occurrences, just one alignment between 5 and 10 percentage occurrence for Dd, with all others falling below the 5% mark. Generally, NON-engineering seniors spent more cognitive effort engaged in Cp and An than the pre-engineering seniors.

Comparisons of the pre-engineering (ENG) and non-pre-engineering (NON) group data using a t-test revealed significant differences which existed only for consulting information about the problem (Cp). Yet, findings indicate that differences existed near the threshold for proposing a solution (Ps), postponing a design action (Pd), and looking back (Lb). Consequently, the NON group spent statistically significant more cognitive effort consulting information about the problem (Cp) than the ENG, while they exerted considerably less effort proposing a solution (Ps), postponing a design action (Pd), and looking back (Lb) than the ENG group. Thus, the low cognitive effort of many of the design issues for coding scheme two (e.g., Pp, Pd, Ca, etc.), coupled with large standard deviation, may be of little substance. In turn, the statistics are not robust enough to be used statistically for comparison.

Improving the Teaching of Design Cognition

One of the more significant findings from the research is that the instructional and pedagogical approaches applied to the engineering students do not appear to significantly prepare them to any greater degree than students who do not take pre-engineering-related coursework. As a result, educators should intentionally and purposefully present strategies and heuristics for students, improve feedback through assessments (pre, formative, and summative), and more deliberately consider how design challenges can be used to situate students within a context that intentionally and appropriately develops their cognitive abilities.

Curriculum Planning

As technology, design, and engineering educators plan curriculum development, they can consider the findings from this research to target areas in which students appear to struggle (e.g., problem identification at the system level, performing calculations to analyze a proposed solution, retracting a previous design decision). For example, future curriculum could be designed with the end in mind, by ensuring that students are provided ample opportunities to not only understand the design process but also holistically and at a detailed level. Specifically, where students, as beginner designers, appear to struggle educators can provide additional curricular time, for students, to reflect and review the process they went through, since often, students apply the design process through a one cycle approach rather than having the opportunity to review their thinking processes, and subsequently become better thinkers, and thus improve their metacognition. Moreover, as curriculum is being developed, teachers should ensure time is spent on the use of materials and tools, to understand how things work and function, since students exerted less cognitive effort toward the actual functionality of the design solution and, overall, the functionality of the system as a whole. One recommendation is that this should not be a stand-alone opportunity or just assumed. Rather, providing targeted learning outcomes, as well as ongoing opportunities to explore materials, hardware, fasteners, and tools, should be deliberately infused into the curriculum, to ensure students navigate design and can activate their prior knowledge and produce improved designs.

Teaching Methods and Assessment for Learning

The conclusions drawn from each of the research questions of this study have implications for the instruction of technological/engineering design-based teaching and learning and preparation of students equipped to effectively use engineering design to solve authentic problems. The findings indicate minimal differences existed between students who take pre-engineering and students who do not, which questions the effectiveness of pre-engineering programs. However, the findings and implications are limited to the participants that were selected from three high schools in Southwest, Virginia. Thus, the implications presented in the following section should be reflected on in terms of the context of each school setting.

Based on the results of this study, a plausible implication is that technology and engineering education should provide students' explicit instruction and strategies on design issues within the *analysis of the problem* domain since students from both groups spent minimal effort engaged in these cognitive activities in contrast to experts. Merely demonstrating to them how they can make an evaluation related to the problem (Ep) and how or when to postpone analysis of the problem (Pp) may substantially increase their decomposition of what the problem encompasses. Considering that this is an effective skillset of expert engineers, intentionally preparing students how to analyze the problem and consult information about the problem will better equip them in practices of engineering design. In turn, this will afford them the ability to meet the fundamental goal of using engineering design, which is to possess the skillset necessary to solve authentic problems (Mentzer et al. 2015).

This study suggests that some teachers' instructional practices and pedagogical approaches may unintentionally guide students toward some cognitive processes and not to others. Considering the research which suggests "that students who are asked to think learn better" (Brookhart 2014, p. 5), this study may aid teachers in recognizing students' natural cognitive ability as they enter a classroom environment. Furthermore, findings can also assist educators in identifying which cognitive processes students may consider more useful than others when designing and therefore offer insight into which cognitive processes will not be employed. This data could be used in the design of specific learning tasks to evoke particular cognitive competencies.

For example, teachers could employ a pre-assessment or even use the findings from this study as a baseline. This understanding would afford teachers the opportunity to choose strategies and examples of how to attend to other cognitive processes that do not receive adequate attention. For instance, identification of where students generally exert more and less cognitive effort can support educators in guiding students toward improved design cognition. This recommendation for instructional practice is also in line with Strimel (2014), who recommends that utilization of the most common or least commonly used cognitive processes students employ can serve as "indicators of voids in curricula, instruction, and student learning" (Strimel 2014, p. 161).

Additionally, this research also seems to indicate that if teachers did not ensure that specific skillsets were mastered, then they would not be implemented by the students. Thus, if students have not developed a level of competency within a specific domain problem, whether that be for design skills, fabrication, or STEM content knowledge, they will not be able to integrate such skillsets effectively. This is comparable to sports training. Athletes do not go from drills, which are focused on specific skills, to an official game. Rather, they must have practices and scrimmages, perhaps prior to a final performance task. In the example used in this situation, students had not yet mastered the individual skills to complete a design solution. Thus, ensuring that students are given ample time in each of the design phases, to break down the skill and master it, will enable them to think more productively when engaged in solving a design problem.

Another example of an instructional tool that can provide timely and specific feedback for students to navigate the design process is for teachers to use the informed design teaching and learning matrix (Crismond and Adams 2012). This tool can easily be used by teachers as students navigate the design process, to ensure they are applying nine engineering design strategies and associated patterns, which compare novice and informed design behaviors, which are also linked to learning outcomes and instructional practices. For example, the matrix can be used by teachers to direct instruction and assessment to ensure students navigate across the beginner to informed design continuum for their ongoing or recently completed design project. Also, students can rate themselves for where they are in the process or have their peers evaluate their work. The following nine design strategies can be examined (understand the challenge, build knowledge, generate ideas, represent ideas, weigh options and make decisions, conduct experiments, troubleshoot, revise/iterate, reflect on process). Then, each design strategy is linked with specific patterns that then contrast a beginning designer and an informed designer. For example, the design strategy, *understand the challenge,* includes the pattern, *problem solving* (beginner designer) vs. *problem framing* (informed designer). Within the matrix, examples of learning goals and strategies are included for instructors/students to reflect, evaluate, and improve their ability to successfully design. The prompts can be used in both a formative and summative manner throughout the design process, with additional scaffolding included.

Additional Related Areas for Review

There are a number of resources available that support the effective instruction of design cognition that could facilitate students focusing on all areas of the design process, for example, having students exert more effort to brainstorming and ideation through the use of design heuristics. Teachers can use design heuristics cards to ensure students are thinking creatively and ideating effectively, as they use one, or many of the 77 specific strategies researchers have identified to help them generate novel designs that are different from each other, which can lead to innovative concepts. Another area of research is to focus on how students gather information, since this was also an issue that students faced. For example, Mentzer (2014) reported that students struggle with recognizing what information is most useful in facilitating a quality solution. Additionally, students background knowledge, memories, and experiences can influence how they think and how they navigate a design problem and solution (Mentzer 2014). Subsequently, awareness of how students approach a problem can aid teachers in scaffolding experiences for students to think through and solve a design task successfully.

Additional Recommendations for Successful Teaching of Design Cognition

One area that has not been investigated in this study due to time and task constraints is the students' cognitive issues surrounding modeling, prototyping, and use of materials. Though this relates directly to the design task recommendations previously discussed, future research should examine and investigate how students model, prototype, and use materials. This recommendation considers findings from Strimel (2014), who allowed students the opportunity to engage in such experiences. He found that the design solutions students proposed were impacted by the materials they had access to and were familiar with in the classroom. In turn, the mere introduction of these components, which are common in truly authentic design tasks, may affect students' distribution of other cognitive issues. Finding opportunities to assess the differences between specific approaches to immerse students in modeling, prototyping, and use of materials can enhance the delivery of design and improve student's thinking capability.

Another area is to break down the design task or compare the structure of design problems and the effect it has on students solving of design issues. Even the scaffolding of more complex problems over time will help to develop students thinking ability. As reported in the literature review of this study, multiple problem types exist, each warranting a more or less constrained approach that may affect student's navigation of solving the design task and result in alternate distribution of cognitive processes. Furthermore, inclusion of criteria and constraints that are related to scientific and mathematical principles may influence students' distribution of their cognitive effort. Future studies would benefit from aligning design challenges to students' curricula selection and background experiences.

Lastly, students' discussion should be examined to distinguish differences that may not have been discovered from frequency counts and percentage occurrences. For example, the findings that depict where students spend the majority of their cognitive effort would benefit from a content analysis of how they are analyzing the problem, thinking about systems, and using mathematical calculations. One group of students therefore may be characterized as using simple addition and subtraction calculations, while another group of students may have applied Pythagorean theorem to solve a design issue. These types of issues could not only further distinguish difference between groups but also illustrate additional details of how students navigate engineering design.

Appendices

Appendix 1: Problem Domain: Degree of Abstraction

Category	Code	Definition	Example utterances
System	0	The designer is considering the problem from the point of view of the user	"OK let's just see so I've got an idea of what this backpack design looks like"
Subsystems	1	The designer is considering the problem in terms of the subsystems	"Side mounted alright"
Detail	2	The designer is considering the details of the subsystems	"What we're going to do is we're going to run these stays to em to this er this fitting here"

Note. Adopted from: Purcell et al. (1996)

Appendix 2: Strategy Classification Scheme

Category	Code	Definition	Example utterance
Analysis problem			
Analyzing the problem	Ap	The designer is analyzing the problem	"What is the system going to need to do…"
Consulting information about the problem	Cp	The designer is referencing external information they were provided	As above but using external information
Evaluating the problem	Ep	The designer is making an evaluation related to the problem	"That's an important feature…."

Category	Code	Definition	Example utterance
Postponing the analysis of the problem	Pp	The designer decides to postpone analysis of the problem	I can find that out later….."
Synthesizing solution			
Proposing a solution	Ps	The designer presents a proposal for solution	"The way to solve that is"
Clarifying a solution	Cl	The designer provides clarify for a solution they have proposed	"I'll do that a bit neater"
Retracting a previous design decision	Re	The designer suggests a previous decision should be retracted	"That approach is no good what if I…."
Making a design decision	Dd	The designer makes a decision between multiple solution proposals	"Ok. We'll go for that one…"
Consulting external information for ideas	Co	The designer consults external information for ideas related to the solution	"What are my options…."
Postponing a design action	Pd	The designer decides to postpone analysis related to a design action	"I need to do…later"
Looking ahead	La	The designer anticipates what will need to be completed ahead of time	"These things will be trivial (difficult) to do"
Looking back	Lb	The designer reviews past decisions	"Can I improve this solution?" –"do I need all these features"
Evaluating solution			
Justifying a proposed solution	Ju	The designer decides to postpone analysis of the problem	"This is the way to go because…."
Analyzing a proposed solution	An	The designer analyzes a proposed solution	"That will work like this…"
Postponing an analysis action	Pa	The designer postpones the analysis of an evaluation	"I'll need to work that out later"
Performing calculations to analyze a proposed solution	Ca	The designer performs calculations related to an evaluation of a proposed solution	"That's seven inches time three"
Evaluating a proposed solution	Ev	The designer evaluates a solution	"This is faster, cheaper etc…."
Explicit strategies			
Explicitly referring to application knowledge	Ka	The designer references knowledge specifically related to a final design	"In this environment it will need to be…"
Explicitly referring to domain knowledge	Kd	The designer references domain specific knowledge	"I know that these components are…"
Explicitly referring to design strategy	Ds	The designer references knowledge related to strategies that they are aware of	"I'm doing this the hard way…."

Note. Adopted from: Purcell et al. (1996)

References

Barak, M., & Hacker, M. (2011). Learning theories for engineering and technology education. In M. Barak & M. Hacker (Eds.), *Fostering human development through engineering and technology education* (pp. vii–vxi). Rotterdam: Sense Publishers.

Bransford, J. D., & Vye, N. J. (1989). A perspective on cognitive research and its implications for instruction. In L. Resnick & L. E. Klopfer (Eds.), *Toward the thinking curriculum: Current cognitive research* (pp. 173–205). Alexandria: ASCD.

Brookhart, S. M. (2014). *How to design questions and tasks to assess student thinking*. Alexandria: ASCD.

Creswell, J. W. (2013). *Research design: Qualitative, quantitative, and mixed methods approaches*. Los Angeles: Sage.

Crismond, D. P., & Adams, R. S. (2012). The informed design teaching & learning matrix. *Journal of Engineering Education, 101*(4), 738–797.

Grubbs, M. E. (2016). *Further characterization of high school pre- and non-engineering students' cognitive activity during engineering design*. Doctoral dissertation. ProQuest dissertations and theses database. (UMI No. 3662376).

Hynes, M. M. (2012). Middle-school teachers' understanding and teaching of the engineering design process: A look at subject matter and pedagogical content knowledge. *International Journal of Technology and Design Education, 22*(3), 345–360.

Lammi, M., & Becker, K. (2013). Engineering design thinking. *Journal of Technology Education, 24*(2), 55–77.

Mentzer, N. (2014). High school student information access and engineering design performance. *Journal of Pre-College Engineering Education Research (J-PEER), 4*(1), 4.

Mentzer, N., Becker, K., & Sutton, M. (2015). Engineering design thinking: High school students' performance and knowledge. *Journal of Engineering Education, 104*(4), 417–432.

NAE (National Academy of Engineering) & NRC (National Research Council). (2009). *Engineering in K–12 education: Understanding the status and improving the prospects*. Washington, DC: National Academies Press.

National Science Foundation. [NSF]. (2012). *Collaborative research: Understanding high school pre-engineering student design cognition, comparisons with engineering students*. (Award No. 1160345). Retrieved from http://www.nsf.gov/awardsearch/showAward?AWD_ID=1160345&HistoricalAwards=false

Petrina, S. (2010). Cognitive science. In P. A. Reed & J. E. LaPorte (Eds.), *Research in technology education* (pp. 136–151). Reston: Council on Technology Teacher Education.

Purcell, A. T., Gero, J. S., Edwards, H., & McNeill, T. (1996). The data in design protocols: The issue of data coding, data analysis in the development of models of the design process. In N. Cross, H. Christiaans, & K. Dorst (Eds.), *Analysing design activity* (pp. 225–252). Chichester: John Wiley.

Razzouk, R., & Shute, V. (2012). What is design thinking and why is it important? *Review of Educational Research, 82*(3), 330–348.

Ritz, J., & Martin, G. (2012). Research needs for technology education: An international perspective. *International Journal of Technology and Design Education, 23*(3), 767–783.

Strimel, G. J. (2014). *Engineering design: A cognitive process approach*. Doctoral dissertation. Retrieved from Proquest. (3662376).

Wells, J., Lammi, M., Grubbs, M., Gero, J., Paretti, M., & Williams, C. (2014). *Design cognition of high school students: Initial comparison of those with and without pre-engineering experiences*. Paper presented at the Eighth Biennial international technology education research conference, Sydney, Australia.

Wells, J., Lammi, M., Grubbs, M., Gero, J., Paretti, M., & Williams, C. (2016). Design cognition of high school students: Initial comparison of those with and without pre-engineering experiences. *Journal of Technology Education, 27*, 78.

Zuga, K. (2004). Improving technology education research on cognition. *Journal of Industrial Teacher Education, 14*(1), 79–87.

Part V
Girls and Technology

Chapter 14
Practical Approaches to Increase Girls' Interest in Technology Education

Sonja Niiranen

Abstract Education has an important impact on preparing children and young adults to participate in future society. To introduce a more equitable gender balance in technology-oriented fields and, consequently, in the labour market, our knowledge of technology education and gender-related issues should continue to expand and to receive attention. Numerous studies have indicated the importance of how technological activities are conducted in class and how teachers can influence students' motivation through the application of different pedagogical approaches. This chapter draws on research to enrich understanding of how to increase girls' interest in technology education by highlighting some practical suggestions for developing technology education in the future.

The Questions I Asked and Why They Are Important

In today's society, technology is playing an increasingly important role, and therefore knowledge and abilities learned through the science, technology and engineering lessons become vital for all citizens (Ardies 2015; Banks and Barlex 2014; Ritz and Fan 2015). Thus, engagement with technology is an unavoidably central component of people's lives, and the experiences with technology have an impact on personal interests, career aspirations and social role patterns related to technology (Volk 2007; Williams 2009). A concern has been expressed for many years that there are relatively few women who have entered occupations in the natural sciences, engineering and technology (e.g. Klapwijk and Rommes 2009; Sander 2012; European Commission 2013b, 2016).

Technology education contributes to this challenge in different ways by providing people with required skills and technological literacy, which they need to understand and utilise to become empowered citizens of tomorrow (Compton 2011). It

S. Niiranen (✉)
Tampere University of Technology, Tampere, Finland
e-mail: sonja.niiranen@tut.fi

P J. Williams, D. Barlex (eds.), *Explorations in Technology Education Research*, Contemporary Issues in Technology Education,
https://doi.org/10.1007/978-981-13-3010-0_14

can also provide pupils with participatory opportunities and concrete representations of activity. It has been suggested that problem-based activities can assist people to become critically literate to address issues through active engagement in both tool-related hands-on and discursive practices of technology (Wilkinson and Bencze 2011). The community of technology education clearly has the potential to foster designerly thinking and technological literacy in ways that respond equitably to human needs now and into the future and to work towards sustaining the ecological resources and environments upon which such developments depend (Rockstroh 2013). Technology education would be relevant to the degree that it inspires the creativity of young people to invent what amounts to a 'new', more sustainable world (Elshof 2011; Pavlova 2009).

Thus, it can be construed that in order to introduce a more equitable gender balance in technology education and consequently in the labour market, attention should be focused on offering girls more opportunities to become familiar with technology and gain skills and experience in this area. Therefore, it is relevant to continue to expand our knowledge of technology education and to give attention to gender differences already evident in primary education. In embarking on my research, my purpose was to gain a deeper understanding of what could and should be done to increase girls' access to and interest in technology education in basic education and ultimately women's interests towards technology-related careers. To do so, my aim was to provide information about two questions: (1) How can girls' access to and motivations in technology education be increased in basic education? and (2) What affects women's interests in entering technology-related careers? What both of these perspectives have in common is whether increasing girls' access to and interest in technology education at the basic education level might ultimately increase the number of women who enter technology-oriented fields.

How I Tried to Answer the Questions

In order to investigate those issues, I compiled four sub-studies, with their own specific aims and research questions. Study 1 aimed to define technology education, with a special focus on describing Finnish technology education in detail based on Finland's National Core Curriculum for Basic Education 2004 (that time in effect). I performed a qualitative, theory-driven content analysis which was the best method for describing the meanings of qualitative material in a systematic way due to the use of predetermined analytical criteria. The advantage of using it was that the procedure provided categories to centre the analysis. In the analysis, I categorised the detailed textual descriptions of the objectives and contents of each subject and cross-curricular theme in the NCCBE 2004 into the levels 1–3. Then, when my aim was to produce broad knowledge on primary school-aged students' motivations and interests in technology education, Study 2, an empirical study, was performed to identify students' motivation in primary education. To do so, I conducted a quantitative questionnaire study for students in the fifth and sixth grade (n = 281, 144 girls and 137 boys). In order to examine the structure of pupils' motivations towards technology education, I performed an Exploratory Factor Analysis (EFA) on the motive statements. After extracting the factors using reliability-measurement-counted EFA, mean scores and Cronbach's alphas, the reliability coefficients for the mean scores were computed using the obtained factor structure as a basis. Then independent samples *t*-tests were used to examine gender-related differences in pupils' motives. Also, Cohen's *d* was used to indicate the magnitude of the difference between girls' and boys' means.

In Study 3, also an empirical study, the focus was broadened to include adults and women who had entered a career in a technology-oriented field. The aim of the study was to explore the main factors that had influenced women's decisions to study and enter a career in a technology-oriented field and to investigate whether studying craft, and especially technical craft, during basic education has influenced their decisions. I carried out the study by use of a semi-structured questionnaire. The study group consisted of 12 female technical craft and technology education teachers who had graduated from various universities in Finland and 12 female engineering students from the Tampere University of Technology and Aalto University School of Engineering. The rationale behind choosing participants from these different areas of technology was a desire to investigate whether these women shared similar reasons for entering careers in technology-oriented fields. I analysed the data by using a qualitative theory-oriented content analysis. In the analysis, meaningful sentences or themes and manifest content were chosen as the analysis units. After coding, the analysis units were grouped and categorised based on the higher-order heading of the theory of career anchors (Schein 1996).

Study 4 also focused on women in a technology-oriented field. The aim of the study was to explore inequality that women may experience when studying and working in technical craft and technology education. Specifically, this study focused on investigating the gendered processes that might exist in craft education, especially in relation to technical craft. The study was carried out using semi-structured theme interviews. Potential participants were asked whether they wanted to participate in the study, and interviews were carried out with those who volunteered. The data consisted of interviews with seven female technical craft and technology education teachers who have graduated from various universities in Finland. These teachers were working in schools for basic education teaching technical craft for students in grades three to nine. In the analysis, I carried out a qualitative theory-oriented thematic analysis through identification, coding, analysis and reporting of patterns within the data.

What I Found Out

The findings of Study 1 suggested that to promote girls in technology education, it is important that they have equal possibilities to discover technological topics and gain self-esteem in the field from primary school. If technology related aspects are studied mainly during technical craft lessons, as they were in Finland, and if students are guided to choose between technical and textile craft studies, it can be construed that technology education reveals a gender-related division. This division means that those girls who study textile craft at school in these grades do not participate in technology-related activities that are part and parcel of technical craft studies. The findings indicated also that technical craft is relevant to the degree that it has the potential to develop students' skills in many ways by enhancing the creativity and innovativeness of young people. The importance of these competences in the field of technology is relevant in many areas of life.

Based on the findings of Study 2, it was clear that there are differences in girls' and boys' motivations towards technology education and crafts. In relation to the content of technology, the study showed that girls were clearly more interested than boys in studying environment-related issues. Girls also enjoyed making useful and decorative artefacts for their homes slightly more than boys. In contrast, boys liked building electronic devices. Interestingly, boys and girls were equally happy about creating their own ideas and realising them. Also, creating an attractive artefact was equally important to them. The study also revealed that boys were more self-confident than girls in learning craft skills. In addition to that, boys were significantly more enthusiastic about craft lessons than girls, and they felt that they could learn new things and that it was fun for them to learn how to use different tools. Boys also felt a bit more excited than girls when doing crafts and strongly disagreed that it was boring. Moreover, it was evident that social interaction between the teacher and the students in class was an important factor for girls. Girls hoped to receive support and encouragement from their teachers and had minor fears of doing something wrong. They also needed to receive encouragement and appreciation for their technical competence. Findings also indicated that in general, students enjoyed working in groups and preferred working with someone.

Whereas the first and second studies aimed to investigate technology education curriculum at schools, and also students' motivation towards it, Study 3 sought to determine the main factors that had an effect on women's decisions to study and enter a career in technology-oriented fields. Additionally, the specific focus was on finding out whether studying crafts during basic education had affected women's decisions in their career aspirations. Concerning women in technology-oriented fields, it was evident that the most influential career anchors were their high level of competence and familiarity of the field. With technology education teachers, strong self-confidence in crafts was addressed in this context, and with engineering students, particularly skills in mathematics, but also physics and chemistry, were highly important elements. Engineering students stated their awareness of how good skills in mathematics are a tool for a wide range of pathways in higher technical education. Also, a highly influential factor for these women in choosing their career seemed to be familiarity with the field or the examples of, and encouragement from, their families.

In Study 4 I focused on identifying the inequality that women may experience, when studying and working in a technology-oriented field. Specifically, the study focused on investigating the gendered processes (Acker 1990) that might exist in craft, especially in relation to technical craft, as being a representative part of technology education in basic education. The study showed that six of the seven women had experienced gendered patterns, when choosing which craft (textile or technical) to study or some of them did not even have any opportunity for that decision. While they had studied textile craft, many of their reasons for choosing revealed aspects of allowed behaviours or institutionalised means of maintaining the division in craft. Their reflections were:

T1: 'I chose textile craft because I felt that it was the way it should be done; however, I also liked textiles a lot'.
T2: 'The atmosphere then was that technical craft was for the boys and something else was for the girls'.
T3: 'I would have needed some encouragement or a friend with me to choose technical craft'.
T4: 'Girls and boys were separate, girls in textile and boys in technical craft'.
T5: 'I did not get much help or encouragement from the technical craft teacher, so I chose textile craft because it was easier for me'.
T6: 'I wanted to choose technical craft, but I was told at home to choose textile craft'.
T7: 'At that time, there was not any decision making about this question'.

It could be seen that many of these women chose textile craft instead of technical craft in primary school due to a tacit assumption that girls should automatically choose textile craft or based on other reasons such as parents' encouragement or peers' decisions or group pressure.

Six of the participants remembered having only male technical craft teachers during their basic education (grades one to nine). This result revealed that the image of technology as a masculine domain has been striking, but in addition, how working was pedagogically organised and what students did during lessons influenced girls' perceptions. Some of the participants remembered that the products they were guided towards during technical craft lessons were gendered for female students, for example, a doll's bed, and that almost all the products were predesigned by a teacher (male) and therefore they were perceived to have a male perspective for using them. Some of the women also remembered gendered actions by their teachers, such as never receiving help at all from the teacher during the lesson or the teacher's unwillingness to help them solve problems or show them how to do something. One of the participants reported that it was only the teacher who could use the machines, while they as students (girls only) used hand tools.

All seven participants had experienced gendered patterns involving the enactment of dominance, submission, questioning or wondering from male teachers, colleagues, technical support staff at school or boys at school during their studies or when working as a teacher. This set of processes were further divided into three subcategories:

1. Belittling and questioning: that describes a situation where a person speaks to another in a way that patronises or belittles the other person on the basis of

gender by using questions such as the following: 'Oh my, do you really know how to do this?', 'Do you actually know what this is?', 'Well that should be done this way, you know' or 'Well you don't need it anyway, so I don't have to show you that'.
2. A request to prove skills: that describes a scenario where a woman is asked to prove her skills, for example, 'If you can't prove that you are adequately skilled and really able to do this…,' or a scenario where someone is looking for specific qualifications but gets 'angry' because a person is qualified but is a woman. In this context, however, some of the participants experienced women being used as a good example of a technology teacher based on their superior skills.
3. Denial: that describes the behaviour of a person who will not cooperate at all or will not accept a woman as a colleague without receiving an extra compensation.

Six of the participants presented the aspects or assumptions of a woman's technical craft identity as a member of that group. The most evident assumption was related to the expectation of having excellent technical skills. Also, possessing traits of masculinity such as being relaxed and not taking things too seriously was mentioned in one participants' response as she expressed that 'I am, myself, quite relaxed and do not stress easily and I also do not want to be with people who take things too seriously. I felt that male students are not like that and knew that many of them were going to study technical craft, so I thought that studying with them would be nice'. Also, one participant said 'often female students were working with a male student to get help and support, but I did not have one to work with. – I wanted to show that I can do it alone and manage without male help'.

How This Research Can Be Used to Improve Teaching and Learning

Curriculum Planning

One of the evident challenges related to technology education is that in some countries it does not have a discrete status in basic education, and therefore it is not considered to be an independent subject. Thus its aims are very general in nature or, on the other hand, very subject related, i.e. technology is taught during the craft lessons. Even though this looseness and ambiguity is often well-meant in its desire to give room for teachers to adapt their instructional practice to the specific pedagogical circumstances, it might have an effect on the gap between curriculum guidelines and actual teaching (Rasinen et al. 2009). Without the guidelines of a curriculum and clear operational instructions on how to implement that curriculum in practice, particularly if there is no easy access to resource materials, technology education might not get its due time in the busy school environment. The importance of competencies in the field of technology for several areas of life should be emphasised, but who will do that in the future? This reveals a challenge for teacher education institutions and also school teachers. Therefore, it is necessary for curriculum writers and teacher educators to pay more attention to these challenges by providing a wider range of courses on technology education, particularly at the teacher education institutions. They should also take care that all students would have an equal access to choose technology courses or even make some courses compulsory for all students, particularly for those who are studying to become teachers in the primary schools.

A deeply gendered history of technology studies and the discourses relating to gender in technological activity have labelled technology as a masculine and male-dominated arena. Thus, one challenge related to technology education seems to be gender-related divisions between studies in some countries as well as gendered entry trends in technology education (Murphy 2007; NCCA 2017; Virtanen et al. 2015). Girls seem to mainly study textile crafts/technology, while boys focus on technical crafts such as resistant materials (woodwork and metalwork) or electronics. One might ask whether girls need encouragement to pursue a wider range of technical subjects, rather than those defined by the role of a traditional homemaker. And what about boys? Is it not equally important for boys to learn skills that are needed when using soft materials? As relevant to the implementation of technology education, the experience of children with technology is a matter of equal opportunities: all children should have the opportunity to gain technological literacy. Teachers play a key role in dismantling gendered practices and renewing the image of technology education because they are well placed to alter students' perceptions and indeed their whole identity (Murphy 2007). As technology educators, we should provide our students equal possibilities to discover technological topics and gain self-esteem in the field.

Students' Motivation Towards Technology Education

The nature of technology education provides students with a systematic approach to solving problems and a context in which students can test their own knowledge and apply it to practical problems. During activities, designers are involved with continual reflection, brainstorming and prototyping and learning by iteration and from feedback and failure and by noticing and troubleshooting, in a dialogue with ideas, material and people (Adams et al. 2003; Crismond and Adams 2012). To help students to develop a positive attitude about and relationship to technology, educational practices should include a broad view of technology and embrace future-oriented conceptions of technology and related careers (Murphy 2007, 250).

In relation to the students' motivation towards technology, education findings evidenced that there are differences in girls' and boys' motivations concerning the contents of technology education and how important students perceive the subject to be. Compared to boys, girls were significantly more interested in studying environment-related issues, inventing solutions for keeping the environment clean and preserving nature. Although the extent to which these contents are related to technology education in students' responses cannot be fully ascertained, this finding indicated that environmental aspects motivate girls. Girls enjoyed also slightly more than boys making useful and decorative artefacts for their homes. In contrast, boys liked building electronic devices. Boys and girls were equally happy about creating their own ideas and that an attractive artefact was important to them. These results indicate that students care about what kinds of artefacts are made in craft lessons and therefore it is important to leave room for students' own design in their projects.

Moreover, it was evident that social interaction between the teacher and the students in class was an important factor, especially for girls. Girls clearly hoped to receive support and encouragement from their teachers and seemed to have minor fears of doing something wrong. They also needed to receive encouragement for and appreciation of their technical competence, particularly from their teachers. Also, a study of autonomy-supportive teachers highlighted that those teachers who supported students' autonomy and initiative create more intrinsic motivation (Reeve et al. 1999). Boys wanted more than girls to solve problems and test and try different things. In addition, boys felt more than girls that they could learn new things and work carried them away when working in a craft lesson. They also believed that it was fun to learn how to use different tools and to use these tools well. Additionally, boys felt that their family encouraged them to do crafts. These findings show the importance of introducing girls to technology education and taking an extra effort in encouraging them in their studies.

If these considerations and findings are taken into account in curriculum planning and teaching, it might increase girls' motivation to study technology-related topics. However, as a conclusion, I wish to note that activities should be planned and presented in such a way that all students would be interested in them and might see technology education as something valuable for them, thereby becoming motivated to study it. Gender-sensitive education should create such learning experiences that recognise and acknowledge girls' and boys' different interests but should, at the same time, strive to encourage children to be individuals.

Encouraging Students for Studying Technology

It has been stated that women's presence in technological fields is essential, because diversity fosters excellence in research and innovation (European Commission 2013a). One aim of schools, as institutions, is to respond to global economic challenges and help students see the breadth of possible study and career options. Concerning the findings on females in technology-oriented fields, it was showed that the women who were studying at the university level and later working in technology-oriented fields did have a high level of competence related to the field they have chosen to study or work in. They had also had examples in, and encouragement from, their families about technological fields, and therefore they have received significant intellectual capital regarding technology or engineering in their childhood, which was influential in their occupational choices.

To introduce more gender balance equality in the labour market, attention should be given to the entire set of factors affecting career choices. This is a question of unused potential. How about those girls who do not have first-hand examples of technological professionals in their families or do not see technology as a relevant topic for them in pursuing technology education and careers? Thus, it is important that schools take on more responsibility for providing students with greater opportunities to obtain practical experience and information about career options in technology-oriented fields. Experienced teachers of technology-related subjects are crucial to support the development of the technology skills and knowledge that students will require in their working lives. Might improved technology education, in the context of career choices of girls, increase girls' interest in technology education and advance women in technology-related careers?

Also, based on the findings, I argue that we must all, as part of the technology education community, reflect on our attitudes regarding gender and how we reflect these attitudes in our speech, gestures, actions and behaviours. Negative communication and encounters should be replaced with positive and encouraging communication that supports the growth and development of students' identity and self-esteem. In order to achieve this, we should understand that there are individual differences as well as group differences between the needs, behaviours and attitudes of girls and boys or women and men. Furthermore, as Murphy (2007) notes, attention should be placed on dismantling assumptions about what girls and boys can and want to do, and students should be offered the support needed to develop new learning habits.

What Else Would Be Good to Know

During this research process, I did become convinced of the impact or potential that technology education might have on advancing women in technology-oriented fields. However, in order to increase girls' interest in technology education as a way to advance women in technology-oriented fields, innovative ways of thinking are needed. It is evident that technology education can provide pupils active engagement and possibilities for hands-on working in relation to technology and that the teachers are the key persons to show and enhance the technological skills and knowledge that are needed later in life. It has also been pointed out that based on recent recognition, a variety of cognitive skills can be developed and nurtured through application to a practical context (Williams 2009). Commonly, craft and technology education or design and technology education emphasises learning by doing or learning while designing. If we recognise learning by doing as a fundamental part of developing students' understanding and tacit knowledge that can support future ventures, should it be seen in a key role in learning? Although the learning by doing approach has not had a central focus in embarking on this research,

I must note that it has been and still is an inherent component in various aspects of Finnish craft and technology education.

Let's assume that teachers would focus on gender-sensitive technology education by creating such learning experiences that encourage children to be individuals, dismantling assumptions about what girls and boys can do and offering all the students support needed to develop new learning habits. Would this provide equal opportunities for all pupils to obtain greater experiences with and information about technology? Also, as a recommendation, by emphasising learning by doing in technology education, many of the findings of these studies could be taken into practical use in teaching. Therefore, I think it is important for technology education teachers to keep on sparking children's interests towards technology by doing things.

References

Acker, J. (1990). Hierarchies, jobs, bodies: A theory of gendered organizations. *Gender & Society, 4*(2), 139–158.

Adams, R. S., Turns, J., & Atman, C. J. (2003). Educating effective engineering designers: The role of reflective practice. *Design Studies, 24*(3), 275–294.

Ardies, J. (2015). *Students' attitudes towards technology. A cross-sectional and longitudinal study in secondary education*. Doctoral dissertation. Antwerpen: Universiteit Antwerpen.

Banks, F., & Barlex, D. (2014). *Teaching STEM in the secondary school: Helping teachers meet the challenge*. New York: Routledge.

Compton, V. (2011). Technology in the primary sector in New Zealand. The journey this far and where to next…. In C. Benson & J. Lunt (Eds.), *International handbook of primary technology education. Reviewing the past twenty years* (pp. 29–38). Rotterdam: Sense Publishers.

Crismond, D. P., & Adams, R. S. (2012). The informed design teaching and learning matrix. *Journal of Engineering Education, 101*(4), 738–797.

Elshof, L. (2011). Technology education: Overcoming the general motors syndrome. In M. de Vries (Ed.), *Positioning technology education in the curriculum* (pp. 145–162). Rotterdam: Sense Publishers.

European Commission. (2013a). *Gendered innovations. How gender analysis contributes to research. Research and innovation.* Report of the expert group 'innovation through gender'. European Commission. Luxenbourg: Publications Office of the European Union.

European Commission. (2013b). *She figures 2012. Gender in research and innovation. Statistics and indicators.* European Commission. Luxenbourg: Publications Office of the European Union.

European Commission. (2016). *She figures 2015. Gender in research and innovation.* European Commission. Luxenbourg: Publications Office of the European Union.

Klapwijk, R., & Rommes, E. (2009). Career orientation of secondary school students (m/f) in the Netherlands. *International Journal of Technology and Design Education, 19*(4), 403–418. https://doi.org/10.1007/s10798-009-9095-7.

Murphy, P. (2007). Gender and pedagogy. In D. Barlex (Ed.), *Design and technology: For the next generation* (pp. 236–251). Shropshire: Cliffeco Communications.

National Core Curriculum for Basic Education (NCCBE). (2004). Helsinki: The Finnish National Board of Education.

National Council for Curriculum and Assessment (NCCA). (2017). *Background paper and brief for the review of junior cycletechnology subjects*. Ireland: NCCA.

Pavlova, M. (2009). Conceptualisation of technology education within the paradigm of sustainable development. *International Journal of Technology and Design Education, 19*(2), 109–132. https://doi.org/10.1007/s10798-008-9078-0.

Rasinen, A., Virtanen, S., Endepohls-Ulpe, M., Ikonen, P., Ebach, J., & Stahl-von Zabern, J. (2009). Technology education for children in primary schools in Finland and Germany: Different school systems, similar problems and how to overcome them. *International Journal of Technology and Design Education, 19*(4), 368–379. https://doi.org/10.1007/s10798-009-9097-5.

Reeve, J., Bolt, E., & Cai, Y. (1999). Autonomy-supportive teachers: How they teach and motivate students. *Journal of Educational Psychology, 91*, 537–548.

Ritz, J. M., & Fan, S.-C. (2015). STEM and technology education: International state-of-the-art. *International Journal of Technology and Design Education, 25*(4), 429–451. https://doi.org/10.1007/s10798-014-9290-z.

Rockstroh, D. (2013). Are we there yet? Questioning whether sustainability is the destination or journey for design & technology education. In P. J. Williams & D. Gedera (Eds.), *Technology education for the future: A play on sustainability* (pp. 400–406). Waikato: University of Waikato.

Sander, E. (2012). Biographies of female scientists in Austria: Results of an interview study. In C. Quaiser-Pohl & M. Endepohls-Ulpe (Eds.), *Women's choices in Europe. Influence of gender on education, occupational career and family development* (pp. 107–122). Münster: Waxmann.

Schein, E. H. (1996). Career anchors revisited: Implications for career development in the 21st century. *Academy of Management Executive, 10*(4), 80–88.

Virtanen, S., Räikkönen, E., & Ikonen, P. (2015). Gender-based motivational differences in technology education. *International Journal of Technology and Design Education, 25*(2), 197–211. https://doi.org/10.1007/s10798-014-9278-8.

Volk, K. S. (2007). Attitudes. In M. de Vries, R. Custer, J. Dakers, & G. Martin (Eds.), *Analyzing best practices in technology education* (pp. 191–202). Rotterdam: Sense Publishers.

Wilkinson, T., & Bencze, L. (2011). With head, hand and hearth: Children address ethical issues of Design in Technology Education. In K. Stables, C. Benson, & M. de Vries (Eds.), *Perspectives on learning in design & technology education. PATT25/CRIPT8*. London: Goldsmiths, University of London.

Williams, P. J. (2009). Technological literacy: A multiliteracies approach for democracy. *International Journal of Technology and Education, 19*(3), 237–254. https://doi.org/10.1007/s10798-007-9046-0.

Chapter 15
Where Are *All* the Students? Factors that Encourage Female Participants in Technology Education

Vicki Knopke

Abstract Females are significantly underrepresented in technology education classrooms in senior secondary schools. Participation rates for female students in Science, Technology, Mathematics and Engineering (STEM) subjects are lower than those of males. Data show that this has a flow on effect on females entering subjects such as Engineering at the tertiary education level. Williams (Intl J Des Educ 7:1–9, 2011) argues that gender issues in technology education have been under-researched. This chapter aims to identify positive influences that encourage female students to participate in technology education. Using a qualitative, feminist ethnographic case study methodology guided by a sociocultural framework, the study aims to demonstrate how classroom teachers may take steps to engage young female learners in technology education.

The Research Questions

The topic was examined through three research questions. The first research question asked – what factors have influenced female student's choices to take technology education classes as part of their senior school pathway? The investigation was undertaken through an examination of the social construction of realities from the individual and the collective group point of view (Bijker et al. 2012), drawing on theories of women's ways of knowing by Belenky et al. (1986).

The second research question asked – how was teaching and learning conducted and approached in selected technology education classrooms? The study investigated Year 11 female students in three secondary schools. In doing so, the ecology of the learning environment, the context of the learning and social interactions were analysed and triangulated from the staff, student and administrators perspectives.

V. Knopke (✉)
Griffith University, Brisbane, Australia
e-mail: vickiknopke@bigpond.com

P J. Williams, D. Barlex (eds.), *Explorations in Technology Education Research*, Contemporary Issues in Technology Education,
https://doi.org/10.1007/978-981-13-3010-0_15

Looking at the three views of the people in the study (the students, the staff and the administrators) helped ensure that the observations and responses to questions that were asked and activities that were observed were confirmed and based on the curriculum within the classroom as well as its intended outcomes.

The third research question asked – what values were addressed in the teaching and learning in specific contexts in technology education? This aspect examined the multifaceted interpretation of values and analysed the engagement of youth and the teaching staff with the concept of values. The aspect of values was important given that the feminist methodology hoped to find if there was a difference between what male and female expectations were in technology learning. Was there a sociological difference in approaches of females to joining these classes and then how they participated? The adjunct to this inquiry was how teachers should encourage female learners in an environment where there are so few.

Belenky et al. (1986) research lead me to the question of how different genders may learn. The dominance theory does not solve the issue of gender difference. Danilova and Pudlowski (2010) argued that one size does not fit all when it comes to technology and engineering studies. The shrinking pipeline (a term coined to refer to the loss of female numbers in engineering studies) could be due to the use of learning styles that attract some participants and not others. This is as relevant to curriculum planners as it is to administrators and teachers in schools.

The questions were important as they shaped the parameters of the research. The aim of this study was to examine the factors that encouraged and facilitated female students to participate and engage in technology education classrooms and, as a consequence, increase participation rates in the long term.

The research questions provided the steps within the study and guided the investigations that I undertook. The three sub-questions were also designed to look at the practical side of what was really happening in the technology education classrooms with respect to female interactions within the technology setting. The questions enabled one to unpack the broader issue of how to increase the participation of female students in technology education. While the issue has been addressed by writers over the past decades, the issue remains one needing further research as numbers of female students have not increased despite efforts of teachers to attract them as participants.

How I Tried to Answer the Questions

The research questions were answered through the use of social constructivist methodology. This qualitative methodology provided the steps which would cater to the feminist stance that was being used to examine the school settings that were being researched.

An ethnographic case study design was chosen as a framework for the investigation. The ethnographic research design aided the study of the behaviour, beliefs, language and how shared patterns of interacting in an educational setting could be developed over time. Ethnography is the systematic study of people and cultures. It is designed to explore cultural phenomena where the researcher observes society from the point of view of the subject of the study. It can also be a process of analysis to which the data is subjected (Pole and Morrison 2003). 'Ethnographic research in this case aimed to use close-up, detailed observations of what was occurring naturally as evidence' (Yin 2003, pp. 1415). The aim is to enhance the understanding of a social action within an educational setting in order to view the wider social and economic context of the class in its location, time and setting. This study was bounded by technology studies, classes and female students in Year 11. This year level is the first of two post-compulsory years of senior high school, prior to university study. The females as a group could be described, analysed and interpreted through patterns within the context of culture-at-work. Ethnography enables in-depth examination of the students in the study. As the researcher I was also the participant observer in the classroom. It was possible to work beside the students and get to know them on a professional level. There were no restrictions on recording or photography or moving about the rooms and talking to both staff and students. This allowed the researcher to interact and be a practical observer, as well as participant, within the technology classrooms which were part of the study.

The research focused on females who were the subject of the study and were identified as the culture-sharing group. The female participants were drawn from within the targeted classes which were Year 11 students who were entering the subject for the first time as part of a 2-year senior studies program. They volunteered to be part of the study and parental permission was gained. Culture by definition covers all human behaviour and beliefs and includes the study of language, rituals, structures, life stages, interactions and communications. The study looked for shared patterns of behaviour and group interactions, similarities, which occurred across the three school sites which were used. The time frame was one teaching unit which generally spanned a term (6–8 teaching weeks) early in the school year.

For each case study the school and environment was described. Data were analysed through four data sets. These included interviews, observation data, audio recordings and photographs taken by the researcher. The interviews included conversations with the female students, the teachers in each of the case studies and the Heads of Department (HOD/administrators).

Data collected through interviews used open-ended questions which were supplied as part of the study and which had received ethical clearance. The interview questions related to the research questions. The schedules were different for the students as compared to the teacher/HOD. Some interviews were individual while others were conducted as a group, but each had a schedule completed and recorded responses where possible.

The sample schedule of questions provided some continuity across the research sites and enabled for a comparison of data when the interviews were transcribed at a later date (Knopke 2015). Not all the participants chose to address every question; however consideration was given to there being enough overlap to enable the researcher to address the three research questions and the seven themes under examination.

Observational data was completed by the participant observer while in the classroom where possible, or shortly after. Overreliance on note making early on in the study meant that interactions were missed so the technique was modified. Photographs were taken and recorders placed as close to participants and field notes referenced and completed soon after the class was finished. Students were asked to complete a repertory grid that related to the artefact they made during the time of the teaching unit which the researcher observed. This was discussed for verification later with the teacher.

An outcome of this type of research is to identify change and actively advocate for change as a result of the findings. The findings were classified into seven themes which emerged as a result of the literature and the pilot study. Data were categorized and systematically analysed in order to answer the research questions in line with the key topic of what could teachers do to further enhance the participation of female students. The seven themes were learning ecology, gender and technology education, language use in classrooms, motivation, role modelling and peer support, sociocultural approaches to learning and values within technology education.

What I Found Out

What I found as a result of the study is summarized in several ways. The first finding related to the methodology used and the second with respect to the content that would address the key research question. More detail will be outlined in relation to how the findings can be directly applied to alter teaching practices in the later part of the chapter.

The focus of this study was on females who identified strongly with one another within the Technology departments of each of the school sites. Despite females from older and younger year levels coming into the workshops, at times the females in the classes I observed did identify as a culture-sharing group and did lend support to one another. These groups varied in size from 2 up to 5 members depending on the timetabling of certain classes in a context of a class of 25 students in total. The awareness of the female learners was not just in the classroom itself but also the educational community in which they operated. The broader community of male design and technology teachers and the professional association were aware of the issues and supportive of any steps to overcome the underrepresentation of gender in this area.

As a result of using a qualitative ethnographic case study approach, the research was able to find some of the voices of the females in the technology classes and analyse the data under the seven themes that developed. Five of the themes emerged from the literature, and two developed as a result of the pilot study which was conducted in one school site.

The analysis of the study was conducted using the seven themes which were:

- Learning ecology
- Gender and technology education

- Language use in classrooms
- Motivation
- Role modelling and peer support
- Sociocultural approaches to learning
- Values within technology education

The structure of the methodology used in this study provides a basis for the replicability of the research in similar studies in the future. There were three research sites, triangulated data from students, teachers and administrators and finally data was analysed and coded via the seven themes listed above. The themes coincided with the sub-questions.

The procedures, processes and instruments used to collect and record the data could be replicated in other research studies. There are a large number of sociocultural factors related to gender and technology in classrooms that need more long-term, in-depth studies.

Each of the seven themes will be outlined, and the following provides a discussion as to what I found as part of the study as they related to the research questions.

Learning Ecology

Learning ecologies that promote learning in technology environments have been described by Siemens (2006) to contain seven elements for knowledge sharing. These elements are flexibility, a tool-rich environment, consistency of practice and time, trust, simplicity, decentralization of learning and a high tolerance for experimentation and failure. Learning ecologies, to Brown (2000), are dynamic, living states. It is a social concept, a practice that is the social construction of reality which stems from humans as social beings acting on their interpretation and knowledge of reality. Learning ecologies can be further analysed as knowledge ecologies: open systems, that are dynamic and interdependent, diverse, partially self-organizing, adaptive and fragile (Brown et al. 1989). A learning ecology is then a collection of overlapping communities of interest, cross pollinating with each other and constantly evolving somewhat analogous to ecosystems such as rituals, response groups, individual class contexts and niches which are managed through the dependent role of members.

One recommendation from the research study was the need to build pedagogical ecologies for technology education. It is suggested that these learning ecologies are based on an awareness of learning styles and values that are unique to females' ways of learning (Knopke 2015). Student backgrounds in terms of socio-economic experiences influence what female students choose to study within school settings. Life experiences and the vocational aspirations of students contribute to student's study plans. In turn their engagement in the learning area shapes what skills the students will develop which in turn shape their contribution to twenty-first century skills. Changing any of the learning ecologies to suit female participants will assist in gaining awareness and participation in the area for all the learners within that system.

Gender and Technology Education

The research uncovered three basic principles stemming from the social constructivist feminist theory of Bijker et al. (1987). Gender as a term is a construct, not created by nature as a result of biology but rather created by and contingent on social and historical processes (Oldenziel 2003). The author acknowledges that power relationships can benefit one group and not another. Social constructivism stems from gender assignment prescribed at birth. Paechter (1998) in her research argues that gender which is usually ascribed to babies at birth shapes our assumptions and expectations about a child's future; however it has more to do with social and cultural values related to that gender than with body features. Secondly, feminist theory can be seen as an ideology as much as a methodology which makes inequalities obvious (Willis et al. 1999). As teachers reflect and discuss issues of gender difference and engagement, the discussion and realization may bring about a change to practice through a reduction of marginalization and listening to the voices of the minorities, notably the female participants.

Language Use in Classrooms

The research reinforced the notion of Spender (1985) that there is a dominance of male language. Men have use and control of language and thus ensure themselves the opportunities to use it from a power perspective. This research could only support the ownership of technology terminology existed because so few females were present to pass on their knowledge of terminology and concepts. The females quickly learned that they needed a mastery of the language to be able to function and compete effectively in the classroom context. For this subject to be socially valued, there is a need to incorporate the values and language of diverse participants. The promotion of a shared language of technology can promote the synergy and harmony between technology and nature exemplified via female language.

Motivation

Motivation is defined in the broadest sense as 'the process whereby goal–directed activity is instigated and sustained' (Pintrich and Schunk 2002). Values, argues Rokeach (1973), have a motivational function: to guide human activity in daily situations; their more long-range function is to give expression to basic human needs. Values' components include motivational, cognitive, affective and behavioural elements.

The research of Zuga (2007) and Wajcman (2004) examined the stigma of artefacts and highlighted the sociotechnical constructivist approaches to teaching females in technology education. Haraway's cyborg-feminists and socialist feminist inquiry was pivotal in exposing the gender blindness of mainstream technoscience studies in order to show the possibilities this area offers females and how they could strategically engage with technoscience within technology (Wajcman 2004).

Artefacts to be made and freedom of choice in the learning process appear to have an effect on the motivation of students as participants in technology education (Thaler and Zorn 2010; Boe et al. 2011). Authors such as Campbell and Jane (2012) have demonstrated that for some students, elements of individual choice have affected their intrinsic motivation. By expanding the amount of internal feedback, their feeling of high levels of autonomy, choice and self-direction, providing an apparent freedom of choice in materials (autonomy), techniques and products to be made, student motivation appears to rise through more active engagement and a willingness to persist. Similarly, Autio (2013) claims self-confidence and expectations for success give value to the options available to females who are studying in technology education today.

In order to bring about change, the approach must be to raise the consciousness of gender and the feminist uses of the construction of ideas and the delivery of programs in technology education. Biological differences between sexes do not determine gender, gender attributes or gender relations. Gender is a constitutive social construction, a social category whose definition makes reference to a broad network of social relations, not anatomical differences (Hacking 1999).

In exploring the perceptions held by students, technology education continues to be perceived as masculine in nature, procedural in delivery and lacking conceptual dimension. Such an enduring perception serves to restrict female interest in the subject (Dakers et al. 2009). Similarly, Klapwijk and Rommes (2009, 406) note the problem with stereotypes: *that 'women' prefer working with people and men* (males) *with things – that if we repeat it often enough it becomes the norm…. Repetition makes it impossible to loosen the unilateral connections….*

The research suggests that motivation can be raised through addressing technology education pedagogy as a positive concept which they (females) come into contact with often and hence will develop skills and knowledge. Frequency of exposure and role models can be the link between technology and femininity (Dakers et al. 2009; Kolmos et al. 2013).

Role Modelling and Peer Support

This theme is closely aligned with motivational factors and communities of interest. It takes into consideration the social factors that teens identify with. Role modelling refers to the individuals who support learners and the guidance given to participants in given situations. Pintrich and Schunk (2002, 384) refer to part of this support as 'peer networks'. These networks are groups with whom students interact in a sociocultural manner. There are further role models of older, respected and familial people in the lives of students. Individuals model themselves on and learn from a range of people. Toren (1996) argues that the human activity in terms of the research relates to the social interactions that can be observed.

Role modelling is the factor which appeared to afford a sense of well-being to female students. Assured support appeared to provide some security of their place when female students commenced their study in a technology class. Increasing female participation would appear to involve a combination of curriculum content, pedagogy and cognitive challenges in a safe and supportive environment. The importance of relationships and acknowledging difference was the key to the plurality of the approach that gained more female participation in the technology education classes.

Sociocultural Approaches to Learning

Sociocultural approaches to learning cover two areas: firstly, the social interactions of learning and secondly, the position and voices of the participants in the study. This study was mindful of the type of critique used by feminist writers and took a positivist view to find the voices of the females in the technology education classes. This perspective addressed the second of the research questions. Sociocultural approaches to learning provide instruction which recognize and empower linguistically and culturally diverse students. Learning in this sense is distributed, interactive and contextual and the result of the learner's participation in a community of practice (Merriam 1998). The collaboration of thinking that results from these processes opens up access to examining thought processes and provides avenues to uncover distinguishing characteristics that could be assigned to gender and could then be addressed by using certain teaching styles and methods ahead of others.

Values Within Technology Education

Values were examined from two aspects: firstly, values and technology education and secondly, values and motivation. Values have been examined within technology education research; however, they appear to be treated as gender neutral. It is the issue of personal values that underlie the feminist perspective, and how these translate into education at the local level which was the focus in this study.

Technology is a human activity requiring a complex understanding of technology, human activity and cultural values that extend beyond the science and engineering of how things work. The Technology for All Americans Project (Dugger 1997) saw the underpinning of technology education as having an ethical and values-based understanding. Technological literacy and goals as noted by Ritz (2009) contribute to this notion. Values in the broad sense of ethical behaviour are those embedded in the Melbourne Declaration as part of the *National values for schooling in Australia* (MCEETYA, Ministerial Council on Education Employment and Youth Affairs 2008).

Research by Pavlova and Turner (2007) demonstrates that sustainable development (SD) learning activities could take place through a conceptualisation of values education. Sustainable development established the cognitive, moral and practical basis for learning, and these factors should be considered during the planning process.

Values, argues Rokeach (1973), have a motivational function to guide human activity in daily situations, with a long-range function to give expression to basic human needs. Values' components include motivational, cognitive, affective and behavioural elements. Instrumental values are motivating because of the attainment of desired end goals, while terminal values are motivating because they represent goals beyond the immediate, biologically urgent goals. They are the conceptual tools we employ to maintain and enhance self-esteem (Rokeach 1973). Terminal and instrumental values are relevant when considering types of behaviour students engage in in classrooms and drive the motivation that individuals employ to achieve their short- or long-term cultural, social or academic goals. In this sense by promoting female language and values, educators are expanding the potential for youth to strive for individual achievements in this sphere of learning.

In Conclusion

This research aimed to contribute to reform processes, encouraging practical and theoretical reflection on the part of the teachers and administrators who were part of the research conversations.

Overall the study found that the impact of curriculum on technology students has an influence on what female students choose to study. Student's backgrounds in terms of socio-economic status largely influence what they choose to study within school settings. Life experiences and vocational aspirations also contribute to female study plans. Having knowledge of the benefits of the subject area, the development of thinking and hands-on skills and the advantages of what technology learning may afford participants appears to make a positive difference to the engagement of female learners.

A female-oriented pedagogy that meets mixed learning styles with structure but provides individuals with independence to problem-solve and discuss issues appears to be an undervalued aspect of technology education programming. The female students who came into the classes had little or no background in any type of technology activity since undertaking compulsory classes in the junior part of high school. What the female students were good at was taking on an understanding of the task at hand, discussing it between their peer group and at times teachers or other adults in the learning environment and then undertaking a project management style of learning to achieve each section of the task. This did not mean that they did each section of the skills – at times some of the females were given outsourcing options, and other male students would offer to plane a piece of wood or cut an angle. The females watched the adept students and then undertook the task, but the teaching staff was wary in that they had encountered female students in the past who had organized others to complete all the tasks to complete an artefact, and they had simply undertaken the assembly and seen it as a successful outcome.

The worth of the output (whether it is a final artefact or a technology activity) and its social and ethical value is part of the criteria that female students use to judge the relevance of the technology education content and courses. Broadening the information base about the subject area will expand its appeal to potential participants.

What the research has demonstrated is that female ways of knowing and doing (Belenky et al. 1986) are different to how many males operate in the technology education.

How might this be used to improve teaching and learning?

The following section looks at how the results of the research may be used to improve the teaching and learning of female students in technology education with respect to curriculum planning, teaching methods, assessment for learning and finally student engagement and participation rates for females.

Curriculum Planning

Curriculum planning is the basis for structuring an inclusive technology curriculum. Support of female teachers and peers appeared to make a difference to the participation of some of the female students in the research study. As the female students began to feel comfortable within the learning environment, they actively engaged in becoming familiar with specific technical terms through discourses conducted at the workbenches. In finding their own identities in the workshops, the female students engaged and participated in the learning to the point where they were able to surpass many of the other students. De Vries et al. (2007), writing on institutional design, noted the need for varied teaching approaches and practices which accommodated differing learning styles, interests, prior knowledge, comfort zones and socialization needs.

Teachers need to re-examine some of their practices in order to entice female participants to participate in technology for the long term. The following explicitly outlines some techniques which could be employed in order to cater to female language and understanding in traditional technology classrooms. These include the use of planning techniques; structured learning styles; management and interactive skills with female students, staff and peers; and freedom to employ one's values to self-motivate for learning.

Techniques used by skilled teachers motivated students in their classes. Of specific appeal to the female students was the pedagogy which provided a structured program that encouraged learning styles which female learners best engaged with. Clear scaffolding provided problem-solving avenues from which flowed project and time management skills and planning that the female learners excelled in. The apparent freedom to learn within what was a sophisticated structured environment did motivate the female students to succeed. This is consistent with Hattie's (2009) claims that student motivation is at its highest when students are competent, have sufficient autonomy, get feedback, set meaningful goals and are affirmed by others. Females will take on internal motivational factors ahead of males who will externalize their academic achievements (Hattie 2009).

The third case study demonstrated that positive learning environments made a positive difference to the female students and their willingness to engage in technology education learning. Feelings of support and the notion of feeling special and unique in a service oriented activity appeared to enhance the engagement of female students. Research by Blackmore (2011) on feminist educational thinking suggests that the discourses of educational change since the 1990s have helped progress the notion of feminism and that the transformation for women in education have brought about changes in social relationships. 'In an era of post masculinity, women need to exhibit strength, strong relations, care and collegiality' (2011, p 208). Lingard and Keddie (2013) in their feminist critiques propose that it is the political agenda which needs to be addressed before the gender divide can be meaningfully changed.

The research of Brizendine (2006) serves to dispute some of these claims but does support the argument that we do need to cater to the learning styles and functionality of females differently to those of male students. Indeed, we need to take into account the cognitive and affective aspects of learning as well as the creative aspects (Knopke 2012, 2015 p.43). This research showed that with some intervention females could find their voice in a technology classroom and use their skills to achieve highly in this context. In teaching what some believed to be a gender-neutral curriculum, male teachers at times missed the potential of the female voices in their thinking skills and organization. In catering to both voices, the whole would be greater than the sum of the individual or divided parts and provide greater appeal.

Teaching Methods

A technology education curriculum, employing language about technology that is intended for all students, needs to incorporate the diversity of people, positions and values in order to reach students and to serve as a socially valued subject in the school curriculum (Zuga 1996). Dakers (2006) argues in a related way that technological literacy and its links to language, values and understanding may provide the criteria and links to teasing out the actions in classrooms and school settings. The research showed that it was the teacher's implementation of language, content and process which contributed to the participation of girls in the classes which were studied.

The study found that a planned female appropriate pedagogy through a structured curriculum has an impact on female students' engagement. Student backgrounds and socio-economic status influenced what they choose to study within school settings. Life experiences and vocational aspirations contributed to the female students' study plans. Having knowledge of the benefits of the subject area, the thinking skills, physical skills and the pathways which the learning may afford participants appeared to make a difference to female choices. These same attributes can be seen in the values orientations which shaped the research questions.

A female-oriented ecology that promotes pedagogy of teaching and learning that mixes learning styles with structured teaching along with the independence to problem-solve and discuss issues has been an undervalued aspect of teaching in technology education. The worth of the output and its social value is part of the criteria that female students use to judge the relevance of courses. More emphasis and focus on values and sustainability should appeal to female participants. Broadening the information base about technology education in the earlier school years will further expand its appeal to potential female participants. The information base could be the provision of an information booklet which provided structure for the project or the course or information on where technology may lead students by way of a career path that was not only for males. Using values such as community contribution, collegiality or sustainability, which appealed to females in projects, was a selling point for the subject.

Assessment for Learning

While this was not a topic which the research focused on, it did emerge during the interviews and class observations with students and teachers. Several points have been noted in relation to gender and technology.

Female participants:

- Want structure in their assessment and will use the teacher provided outlines to guide their work practices.
- Were meticulous in following the criteria and the results of those in the study showed high levels of achievement in the outcomes of both written and practical work over the term they were observed. The criteria provided guidance for achievement but did not stifle the creativity and independent thinking that the female students displayed.
- Were diligent in their paperwork and referred to directions and printed guidance to clarify issues before approaching a teacher.
- Would seek feedback and assurances before proceeding further.
- Achieved outstanding results within the technology classes.

Teachers:

- Believe more females should be undertaking their courses
- Do not believe they are biased in language use and delivery of the technology courses
- Understand the need for structure of assessment and provide this in assessment planning
- Know that female students may require different approaches to teaching – especially in the workshop environment

- Do not believe that they need to alter their practices in any significant manner
- Know that female students who engage in the classes will be high achievers because of their motivational practices

The results have shown that female students will carefully follow the teacher planning, undertake the set activities and follow through with careful presentations and will be the outstanding achievers in their cohort. It is a time factor to gain technological literacy skills which may restrict them and lower their confidence levels when they first enter the technology classes. Teachers need to spend time at the start of a unit or project to ensure that female students have an understanding of the terminologies which will be used on a specific task or project. Teachers did provide technical demonstrations for all students, but this did not ensure engagement or understanding of how to perform that skill until it came up in a particular task. This can be seen as a time factor by some teachers.

Student Engagement and Participation Rates for Females

The female participants in the research study were highly motivated to achieve once they had made the decision to participate in the class. The research looked at whether motivation and values were the first strategy the student's used for selecting to enter the technology classes. The results showed that it was a personal motivation more than an employment motivation that brought about their participation. It was not values related to sustainability. One of the female students in case study 2 felt she had time to make her own decisions at her own pace. She in turn exhibited a confidence and self-assurance over what she was doing and was able to bolster and mentor other students. Belenky et al. (1986) and Ekert and McConnell-Ginet (2003) argue that the emotional control as much as the physical empowerment enables females to be successful.

> *The teacher's expectations are of a high standard of what will be done and despite there being a lot to get through they are really helpful.* (S2) Knopke (2015, 207).

Self-efficacy was the second strategy in engaging and motivating female students in technology education. A belief that one has the capabilities of exercising courses of action to manage situations has been seen as a positive predictor of achievement in task-specific goals and success for females in nontraditional career areas. Cognitive and metacognitive skills focusing on self-efficacy provide motivation to learn. Marra et al. (2009) examined positive outcomes that were achieved with women to understand student satisfaction, achievement and ultimately, retention in engineering programs. Influencing environments in turn sustained persistence and enabled mastery experiences in complex design projects via strategies of instructional demonstrations and encouragement. Positive success led to long-term participation. This is the same factor that the research found for younger students in the secondary context. Students who achieved degrees of mastery of skills became more persistent and resilient to learning within the technology education courses of study becoming leaders within their peer group (Knopke 2015).

The findings send a message to practitioners in the post-compulsory area of senior technology education. A strategy that may assist teachers to revisit and alter the ecology of their classrooms and department to accommodate female participants in technology programs is key to their participation. This research study has shown through current empirical and theoretical research that strategies to promote female participation involve long-term planning, short-term immediate support and constructionist considerations. This stage of schooling is almost too late to gather more support for entry to universities at the career level (Knopke and de la Barra 2013).

The short-term strategies are important, but it is the long-term planning and human resource components (female role modelling) that appear to be making key impacts on female participation and motivation in technology education in early secondary schooling. Role modelling, peer supportive environments, elements of choice and sustainability and the processes to achieve meaningful and valued artefacts are the factors which will bring about further changes. The longer-term strategies are about changing the phenomenon that is socioculturally and psychologically rooted and constructed – 'Women need to be given the explicit message that Technology, in all its aspects, is suitable for women' (Klapwijk and Rommes 2009).

Recommendations for Teachers and Learners

The research indicates that if teachers undertake some of the recommendations provided, then they would be progressing towards a female-oriented pedagogy. This study contributes to knowledge and practice for teachers due to the seven areas of analysis concerning females in technology education.

These comprise of the ecology of classrooms, gender, language use in classrooms, motivation, peer support, sociocultural contexts and values. Female learners, who are the new entrants to the subject of technology education, need to be shown how to adapt to this new learning environment, the language practices that are part of the ecology and the procedures and techniques which again require some depth of learning for technology.

The learning, thinking skill, physical skills development and project management skills enable female students to learn in a relatively new environment. These skills equip them to apply their knowledge to future challenges. Technology departments need to provide time for female learners to develop skills and language that will enable them to function confidently in workshop environments.

A female-oriented pedagogy provides scaffolding problem-solving and skills development through group discussions, collaborations, demonstrations and reasoning in an open context. These factors enable female learners to have choices and make decisions that will motivate them to further engage in technology education.

Teachers need to be mindful that there is no such thing as a 'gender-neutral curriculum'. The making of jewellery boxes instead of tool boxes does not address this issue. Catering to student choices and the reasoning of students provides an understanding and motivation to participate in meaningful projects. Some may be to benefit the community while others have personal meanings. Understanding the technical language quickly, once females enter a technology environment, has been seen as a key factor enabling students to engage with projects, materials and tools. Structured learning booklets/guidelines have assisted self-directed females to progress at a fast pace once they understand the concepts they are working with.

The research showed that peer support and shared group goals enabled females to excel in some areas of technology. Peer support was not limited to same gender or role in the class; however it was relevant to the female learners to have a person to discuss issues of the project with.

The gender-neutral claim by teachers fails to enable female learners to use their voice in the technology classroom or contribute to alternative strategies in a learning context. The female learners often knew the answers a teacher posed but failed to use their voice to propose a solution to a problem. The females had the answers and alternative strategies to achieve the outcomes however they perceived themselves as restricted in a male-dominated environment. Females would collaborate, discuss options and move onto testing solutions and often arrive at higher end results that their male counterparts had not achieved. Use of instruction that focuses on females requires collaboration, discussion of options at all transition points in the program and testing solutions that provide more depth of inquiry.

Teaching strategies and practices which require collaboration, discussion of options, testing solutions and alternative thinking patterns and strong time and project management skills are sound strategies for both female and male students. It is not only the females who will benefit but so will their male peers. The prevailing pedagogy should encourage and perhaps require that these factors are included in technology practices.

Finally, values that females engage with can be different to those upheld by male colleagues. The reason one wishes to engage in a project is at times more important and relevant than the skills needed. Females wanted to see the big picture and then work back through the skills needed to achieve the goal. Teacher planning for joint/shared project development that has a social value would appeal to more females and heighten the view of the worth of what they were engaging in. This relevance of projects and skills should become a sales point for such an engaging subject area for younger female students. There is no short-term fix but rather long-term planning on behalf of the teachers to find the social worth and engage in it.

References

Autio, O. (2013). When talent is not enough: Why technologically talented women are not studying technology. *Journal of Technology Education, 24*(2), 14–30.

Belenky, M. F., Clinchy, B. M. V., Goldberger, N. R., & Tarule, J. M. (1986). *Women's ways of knowing: The development of self, voice, and mind.* New York: Harper Collins.

Bijker, W., Hughes, T., & Pinch, T. J. (1987). *The social construction of technological systems*. Cambridge: MIT Press.

Bijker, W. E., Hughes, T. P., Pinch, T., & Douglas, D. G. (2012). *The social construction of technological systems: New directions in the sociology and history of technology*. Cambridge, MA: MIT press.

Blackmore, J. (2011). Leadership in pursuit of purpose: Social, economic, and political transformation. In C. M. Shields (Ed.), *Transformative leadership: A reader* (p. 2136). New York: Peter Lang.

Boe, M. V., Henriksen, E., Lyons, T., & Schreiner, C. (2011). Participation in science and technology: Young people's achievement-related choices in late-modern societies. *Studies in Science Education, 47*(1), 37–72. https://doi.org/10.1080/03057267.2011.549621.

Brizendine, L. (2006). *The female brain*. New York: Random House LLC.

Brown, J. S. (2000). Growing up digital, how the web changes work, education, and the ways people learn. *Change: The Magazine of Higher Learning (New Rochelle, NY), 32*(2), 11–20.

Brown, J. S., Collins, A., & Duguid, P. (1989). Situated cognition and the culture of learning. *Educational Researcher, 18*(1), 32–42.

Campbell, C., & Jane, B. (2012). Motivating children to learn: The role of technology education. *International Journal of Technology and Design Education, 22*(1), 1–11.

Dakers, J. R. (2006). *Defining technological literacy* (p. 352). Retrieved from https://doi.org/10.1057/9781403983053. doi:https://doi.org/10.1057/9781403983053

Dakers, J. R., Dow, W., & McNamee, L. (2009). De-constructing technology's masculinity: Discovering a missing pedagogy in technology education. *International Journal of Technology Design Education, 19*, 381–391. https://doi.org/10.1007/s10798-009-9099-3.

Danilova, E. A., & Pudlowski, Z. J. (2010). Public understanding of engineering: Consequences and solutions. In *1st WIETE annual conference on engineering and technology education Pattaya, Thailand*. Retrieved from http://www.WIETE.com.au/conferences/1st_wiete/5-04-Danilova.pdf

De Vries, M. J., Custer, R., Dakers, J., & Martin, G. (2007). Differentiation in the technology education classroom. In M. J. De Vries, R. Custer, J. Dakers, & G. Martin (Eds.), *Analysing best practices in technology education* (pp. 235–245). Sense: Rotterdam.

Dugger, W. E. (1997). The next step. *Technology Teacher, 56*, 10–19.

Ekert, P., & McConnell-Ginet, S. (2003). *Language and gender*. Cambridge: Cambridge University Press.

Hacking, I. (1999). *The social construction of what?* Cambridge, MA: Harvard University Press.

Hattie, J. (2009). *Visible learning*. Oxon: Routledge.

Klapwijk, R., & Rommes, E. (2009). Career orientation of secondary school students (m/f) in the Netherlands. *International Journal of Technology and Design Education, 19*, 403418. https://doi.org/10.1007/s10798-009-9095-7.

Knopke, V. (2012). *Gender and technology education: Some theoretical implications*. Paper presented at the explorations of best practice in technology in design & engineering education, technology education research conference, Gold Coast, Australia.

Knopke, V. (2015). *Factors that encourage and facilitate female students to participate and engage in technology education*. Nathan: Griffith University.

Knopke, V., & de la Barra, B. L. (2013). *Motivational factors in female senior secondary students: staying and thriving on the technology education pipeline*. Technology education for the future: A play on sustainability.

Kolmos, A., Mejlgaard, N., Haase, S., & Holgaard, J. E. (2013). Motivational factors, gender and engineering education. *European Journal of Engineering Education, 38*(3), 340358.

Lingard, B., & Keddie, A. (2013). Redistribution, recognition and representation: Working against pedagogies of indifference. *Pedagogy, Culture & Society, 21*, 427 Taylor & Francis.

Marra, R. M., Rogers Kelly, A., Shen, D., & Barbara, B. (2009). Women engineering students and self-efficacy: A multi-year, multi-institution study of women engineering student self-efficacy. *Journal of Engineering Education, 98*(January), 11.

MCEETYA, Ministerial Council on Education Employment and Youth Affairs. (2008). *Melbourne declaration on education goals for young Australians*. Canberra: Ministerial Council on Education Employment and Youth Affairs.

Merriam, S. B. (1998). *Qualitative research and case study applications in education*. Revised and expanded from "case study research in education". Jossey-Bass Publishers, 350 Sansome St, San Francisco, CA 94104.

Oldenziel, R. (2003). Why masculine technologies matter. In N. E. Lerman, R. Oldenziel, & A. Mohun (Eds.), *Gender and technology* (pp. 37–71). Maryland: Johns Hopkins University Press.

Paechter, C. (1998). *Educating the other: Gender, power and schooling*. London: Flamer Press.

Pavlova, M., & Turner, S. (2007). It's never too early: Education for sustainable development, the international journal of environmental, cultural, economic and social sustainability. *Ecological Economics, 48*, 369–384.

Pintrich, P., & Schunk, D. (2002). *Motivation in education, theory, research, and applications* (2nd ed.). Upper Saddle River: Merrill, Prentice-Hall International.

Pole, C., & Morrison, M. (2003). *Ethnography for education*. Buckingham: Open University Press.

Ritz, J. (2009). A new generation of goals for technology education. *Journal of Technology Education, 20*(2), 50–64.

Rokeach, M. (1973). *The nature of human values*. New York: Free Press.

Siemens, G. (2006). *Learning ecology, communities, and networks: Extending the classroom*. Retrieved from http://www.elearnspace.org/Articles/learning_communities.htm

Spender, D. (Ed.). (1985). *Man-made language*. London/Boston: Routledge/Kegan Paul.

Thaler, A., & Zorn, I. (2010). Issues of doing gender and doing technology–music as an 445- innovative theme for technology education. *European Journal of Engineering Education, 35*(4), 454.

Toren, C. (1996). Ethnography: Theoretical background. In J. T. E. Richardson (Ed.), *Handbook of qualitative research methods for psychology and the social sciences* (pp. 102–112). Leicester: The British Psychological Society.

Wajcman, J. (2004). *Technofeminism*. Cambridge: Polity Press.

Williams, J. P. (2011). Research in technology education: Looking back to move forward. *International Journal of Design Education, 7*(May), 1–9. https://doi.org/10.1007/s10798011-9170-8.

Willis, J., Thompson, A., & Sadera, W. (1999). Research on technology and teacher education: Current status and future directions. *ETR & D, 47*(4), 16.

Yin, R. K. (2003). *Case study research, design and methods* (Vol. 5, 3rd ed.). London: Sage.

Zuga, K. F. (1996). Reclaiming the voices of female and elementary school educators in technology education. *Journal of Industrial Teacher Education, 33*(3), 23.

Zuga, K. (2007). *Stem and technology education*. White Paper written for ITEA, 6. Retrieved from https:/iteea.org/mbraonly/library/whitepaper/STEM(Zuga)pdf

Part VI
Information Technology

Chapter 16
Using Technology to Support Discussion in Design and Technology

Adrian O'Connor

Abstract This chapter shares the results of the design and implementation of a conceptual model which uses learning protocols to support communication between teachers and pupils. The findings show that both teachers and pupils responded positively to the integration of new pedagogical and technological approaches into their traditional environments as well as having the capacity to extend teaching and learning beyond the classroom. Pupils embraced this contemporary approach, indicating this stimulated their interest, motivation, and engagement in the regular day-to-day school activity, and the general perception is that it can enhance the processes and conditions of teaching and learning.

The Questions I Asked and Why They Are Important

Despite the pervasiveness of information and communication technology (ICT) in our daily lives, such technology has not been as widely adopted in formal education. When ICT is used within the classroom, its impact on pupil learning is mixed at best. However, technology is linked with increased learning in some contexts, such as when ICT extends teaching and learning and enhances classroom practice (OECD 2015). When ICT supports the pupils' engagement with challenging material, thus extending the learning and enhancing classroom practice, or it helps pupils to assume a greater control over the learning situation by individualising the pace of which new material is introduced or by providing immediate feedback, pupils learn more (OECD 2015). The strongest associations between the use of ICT and enhancing practice are pupil-centred approaches and formative practices for teaching and learning, which include individualised pace and feedback, collaborative learning, and problem-based learning.

A. O'Connor (✉)
Portmarnock School, Dublin, Ireland
e-mail: Adrian.OConnor@ul.ie

P J. Williams, D. Barlex (eds.), *Explorations in Technology Education Research*, Contemporary Issues in Technology Education,
https://doi.org/10.1007/978-981-13-3010-0_16

Although we have not yet become good enough at the kind of pedagogies which make the most of ICT and increase the effectiveness of teaching and learning (OECD 2015), the debate about the use and justification of ICT in schools has been replaced by a discourse of inevitability, where schools are presented as ICT-rich sites of learning. Accordingly, it is argued that supporting discourse is a precursor for the effective use of ICT within teaching and learning, and technology-mediated communication (TMC) provides a medium for supporting discourse between teachers and pupils. However, the space created within this discourse of inevitability enables schools and teachers to integrate technology without any guiding epistemological or pedagogical framework.

As schools and teachers look to integrate systems of ICT in their classroom practice, design and technology (D&T) has continued to be recognised as a potentially rich environment to investigate their role both in teaching and learning (McCormick 2004). This is because the nature of designing in D&T education is a conversational activity (Hamilton 2003, 2004; Hennessy and Murphy 1999; Murphy and Hennessy 2001) that draws on both teacher-pupil and pupil-pupil interactions within the social context. Moreover, D&T has the potential to exploit ICT to such a degree that it can transform both teaching and learning in a way that it has not yet been done (McCormick 2004), as it affords pupils the opportunities for interaction, even when these are not made explicit by the pedagogic stance adopted by the teacher (Murphy and Hennessy 2001). Thus, D&T places conversation at the core of the educational process (Trebell 2007), providing both the context and the opportunities in which ICT can enhance practice through individualised feedback, collaborative learning, and problem-based learning. In the context of D&T, this would include supporting communication with the use of scaffolded interactions where the act of designing becomes the focus of conversation with the interactions being seen as tools to develop mutually appropriated concepts. Hence, the research investigated the development of a conceptual model for enhancing practice in D&T education. Accordingly, the following questions guided the research:

1. What are the perceived effects of supporting discourse using technology-mediated communication on enhancing practice for teaching and learning?
2. How can supporting discourse using technology-mediated communication in D&T education inform the process and the evidence of teaching and learning?

How I Tried to Answer the Questions

To capture the authenticity of the real world of teaching and learning, a *design experiments* approach was identified as being an appropriate strategy (Brown 1992). Though Cohen et al. (2011) acknowledge that a design experiment or *design study* approach does not conform to the requirements of a conventional experiment, it does involve a deliberate and planned intervention, much like a conventional experiment. Brown (1992) suggests that design studies attempt to engineer innovative educational environments and simultaneously conduct experimental studies of those innovations. Accordingly, the data collected should facilitate the generation of *thick descriptions* (Ceertz 1973), particularly of the social processes, cognitive functions, and behaviours in tandem with measuring perceptions in order to understand the dynamics of TMC. A mixed methods approach was therefore chosen for this research intervention as the nature of the analysis would need to focus on both qualitative and quantitative data. This included transcript analysis, pupil focus groups, teacher interviews, and surveys.

The success of any technological initiative is substantially influenced by two major aspects: technical skills and positive attitude (Zhao et al. 2002). Hence, in support of the emerging views expressed by Lei (2009) and Teo (2009) and in an effort to reduce variability among the integration of the technological approach, pre-service teachers were, therefore, most suited for participation within this research, which enabled this research to focus on the development of the pedagogical approach. Although this could suggest an inability to generalise findings to in-service teachers, the findings can inform professional development or training for in-service teachers. Based on a set criteria, seven pre-service teachers (six male, one female) were interested, and suitable, for participation within this research. The pre-service teachers ranged in age from 20 to 27 with a mean age of 22.29 years and a standard deviation of 2.29 years. In collaboration with the pre-service teachers, their cooperating teachers and their associated schools, the 6-week research intervention was integrated into seven classrooms. The participating pupils in this study (n = 104) included 96 male and 8 female pupils. Although the female participants only make up 7.6% of the research sample, this is representative of the gender imbalance in D&T education in Ireland (DES 2007, p.4). The pupils varied in age from 15 to 18 with a mean average of 16.35 and a standard deviation of 0.77, and there was a mean average of 14.86 pupils within each classroom. The EPI Model (Fig. 16.1) was developed as a conceptual framework for supporting discourse (O'Connor et al. 2016, 2018), which provides diagnostic and formative evidence about the quality of pupil learning and efficacy of teacher pedagogy by developing the following elements and processes: the design of the learning environment, the delivery of the educational transaction, and the experiential, procedural, and individual domains of teaching and learning (see below).

The Experiential Domain

The student experience, which has communication at its core, can begin to describe and map learning as evidence-based progression through each stage within the domain (i.e. construct, capture, communicate, cogitate). Cognisant of the direction of new knowledge, which is linked to existing knowledge, where deeper understandings are developed from, and take the place of, previous understandings. The aim of this pedagogical approach is to move pupil understanding along a path of increasingly complex knowledge and skill by focusing on the pupils' readiness to

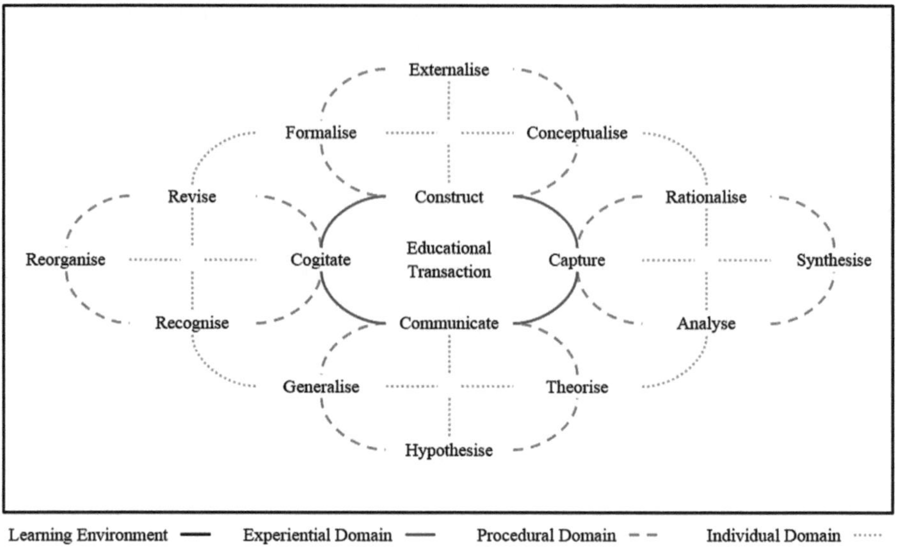

Fig. 16.1 The EPI model

learn and building upon their current stages of understanding. As this process develops through both the pupils' practical activity and social interaction with others, it is the assumption of this research that by integrating the *Experiential Domain*, we can begin to evidence the complex and fundamentally non-linear nature of design-based activity.

In an effort to document evidence-based progress during the educational transaction and practically integrate the proposed EPI Model as a pedagogical approach, teachers and pupils were required to use a virtual learning environment (VLE) in conjunction with mobile technologies (e.g. smartphones, tablets) and stationary computers, to post matters associated with their process of learning and the product of their learning by constructing theoretical knowledge and practical skills, capturing the learning process and evidence of their learning, communicating a/synchronously with group members, and cogitating on their learning process in both collaborative and individual contexts. The structure of posts added to the VLE was defined by each pupil to reflect the experience of their design process and represent the significance of their individual and collaborative approaches to forming solutions and showing evidence of learning. The nature of the data uploaded to the VLE and the modes of representation used were at the discretion of the pupils and facilitated by the file sharing capacity of the VLE, i.e. pupils could attach various text, image, audio or video files and even share links.

The Procedural Domain

Procedural knowledge is the knowledge exercised in the accomplishment of a task and therefore includes knowledge, which, unlike declarative knowledge, cannot be easily articulated, by teachers or pupils seeing as this is typically nonconscious (or tacit). Accordingly, this research classifies the *Procedural Domain* as individual processes of social and cognitive behaviour that evidence knowledge and understanding within a learning environment. In this domain we find a psychological state in which the pupils are trying to make sense of the Experiential Domain, i.e. how pupils construct, capture, communicate, and cogitate evidence of both the process of their learning and the product of their learning when participating in an adaptive educational transaction.

As teachers base their intervention strategies and decisions on evidence (what pupils do, say, make, or write), rather than inference (what pupils know, understand, think, or feel), the capacity to identify pupils' current and targeted progressions is fundamental as to when a person should or shouldn't intervene in the process of pupils' learning. However, to do so properly can prove difficult even for the most experienced teacher. For this to work both in theory and practice, for teachers and pupils alike to base their intervention strategies and decisions on evidence, learning protocols based on a shared language of understanding have to first be established in order to support discourse. Hence, this research proposes that the protocols of the Procedural Domain (Table 16.2) can be used to 'self-code' the posts and comments teachers and pupils add to the VLE.

A unique feature of most VLEs is the hashtagging system which teachers and pupils have the option of integrating during the management of their posts and comments. While uploading their files to the VLE, pupils have the option to create a tag or link between specific pieces of data and the learning protocols of the Procedural Domain. The # symbol, called a hashtag, was used by both teachers and pupils to self-code their individual and collaborative evidence posted as part of the educational transaction. Any one of the learning protocols could be assigned to a post or comment added to the VLE by a teacher or pupil using a hashtag, e.g. #conceptualise. However, it was not mandatory to hashtag as this may cause pupils to tag work for the sake of doing so. This required each pupil to reflect on the data being uploaded and to communicate it in terms of how it related to their own development within the educational transaction. This provides an insight into the perception of how pupils conceive and develop their thoughts, feelings and ideas about the process and product of their learning. It should be noted that the learning protocols were open to interpretation by teachers and pupils and to be used to guide learning and not followed as a chronological set of procedures.

The Individual Domain

Due to the complex and fundamentally non-linear nature of design-based activity, both teachers and pupils can choose to operate within the proposed Experiential Domain and Procedural Domain of the EPI Model in their own idiosyncratic manner. Therefore, a significant contribution that this research makes to the current body of knowledge is the synergy connecting these domains. This is a novel approach to progress monitoring which takes account of the many variables and contexts as the educational transaction develops and changes over time. As a result of the pedagogical and technological approach, this research proposes the *Individual Domain* in response to the physiological and unpredictable irregularity of social and cognitive behaviour. The Individual Domain provides a lens through which to consider the characteristic disposition of knowledge and understanding evidenced in the educational transaction and the multitude of learning styles present in the learning environment by providing the right support and the right feedback, in the right time and place, at the right level.

What I Found Out

Though all pupils (n = 104) in the seven contributing schools volunteered to participate within the study and had initially signed up and created their profiles on the VLE, approximately 84% of all pupils (n = 87) actively took part in the research intervention (working iteratively using the VLE, tagging posts and comments, collaborating with others and uploading subsequent data files (evidence) to complete the learning task). The remaining pupils (n = 17) engaged with the classroom activity as normal practice. Table 16.1 illustrates the percentages of pupils who actively participated within the study and indicates that on average the pupil uptake in each school was approximately 86%.

The process of working iteratively using the VLE by tagging posts and comments with learning protocols, collaborating with others and uploading data files, as evidence, allowed this research to track, manage, and record the complex and fundamentally non-linear nature of teachers' and pupils' activity during the educational transaction. A total of 794 online interactions were recorded using the VLE over the course of the research intervention. These interactions were comprised of teachers' posts (n = 50), teachers' comments (n = 87), pupils' posts (n = 302), and pupils' comments (n = 355). Three hundred sixty-four files were attached by teachers (n = 50) and pupils (n = 314) to individual posts. These files typically included documents, presentations, images, graphics, or videos. The overall word count of all posts and comments by teachers and pupils was 16,542.

Figure 16.2 presents an example of posts and comments uploaded by participants in S_05.

Table 16.1 Participant uptake of research intervention

	Total pupils			Active pupils			
School	*n*	Male	Female	*n*	Male	Female	%
S_01	20	16	4	13	12	1	65
S_02	24	24	0	24	24	0	100
S_03	20	18	2	18	16	2	90
S_04	7	7	0	7	7	0	100
S_05	5	5	0	5	5	0	100
S_06	11	11	0	9	9	0	82
S_07	17	15	2	11	10	1	65
Total	104	96	8	87	83	4	83.65
A.M.	14.86	13.71	1.14	12.43	11.86	0.57	85.93
S. D.	7.24	6.58	1.57	6.63	6.41	0.79	15.89

The Nature of Learning Protocols

As part of the research intervention and with the use of the VLE's hashtag system, both teachers and pupils had the option of creating a link between specific pieces of data and the 12 learning protocols (LP) identified as part of the Procedural Domain. As shown in Table 16.2 below, individual codes (LP01–LP12) and descriptions of each of the learning protocols have been generated for the purposes of coding and analysis.

The emphasis was put on the pupils to identify what protocols they valued as being appropriate to evidence learning with a conscious effort being made by teachers not to impose any criteria or values external to the pupils own experience with the activity. Accordingly, the assignment of learning protocols was at the discretion of the pupils. Table 16.3 illustrates the density at which teachers and pupils assigned each of the 12 learning protocols to both posts and comments throughout the research intervention. As Table 16.3 shows, a total of 413 protocols were assigned to interactions ($n = 794$).

However, of the 794 interactions that were recorded over the duration of the study, 347 of these interactions contained at least 1 protocol, which is approximately 44%. Hence, the remaining 56% of interactions ($n = 447$) had not been assigned protocols. A detailed analysis of the learning protocols assigned by both teachers and pupils reveals that, on average, posts were assigned 0.82 ($SD = 1.04$) protocols by teachers and 0.65 ($SD = 0.66$) protocols by pupils, while, on average, comments were assigned 1.03 ($SD = 0.67$) protocols by teachers and 0.24 ($SD = 0.48$) protocols by the pupils.

As shown in Fig. 16.3, each of the 12 learning protocols were assigned to at least one post or comment by teachers and pupils using a hashtag over the course of the study. This shows that LP10 (Revise) was the most used LP by teachers ($n = 34$), and LP04 (Analyse) was used the most by pupils ($n = 101$), while LP07 (Generalise) was used the least by teachers ($n = 4$), and LP09 (Theorise) was least used by pupils ($n = 2$). The significance lies in the relationship between the frequency of teachers'

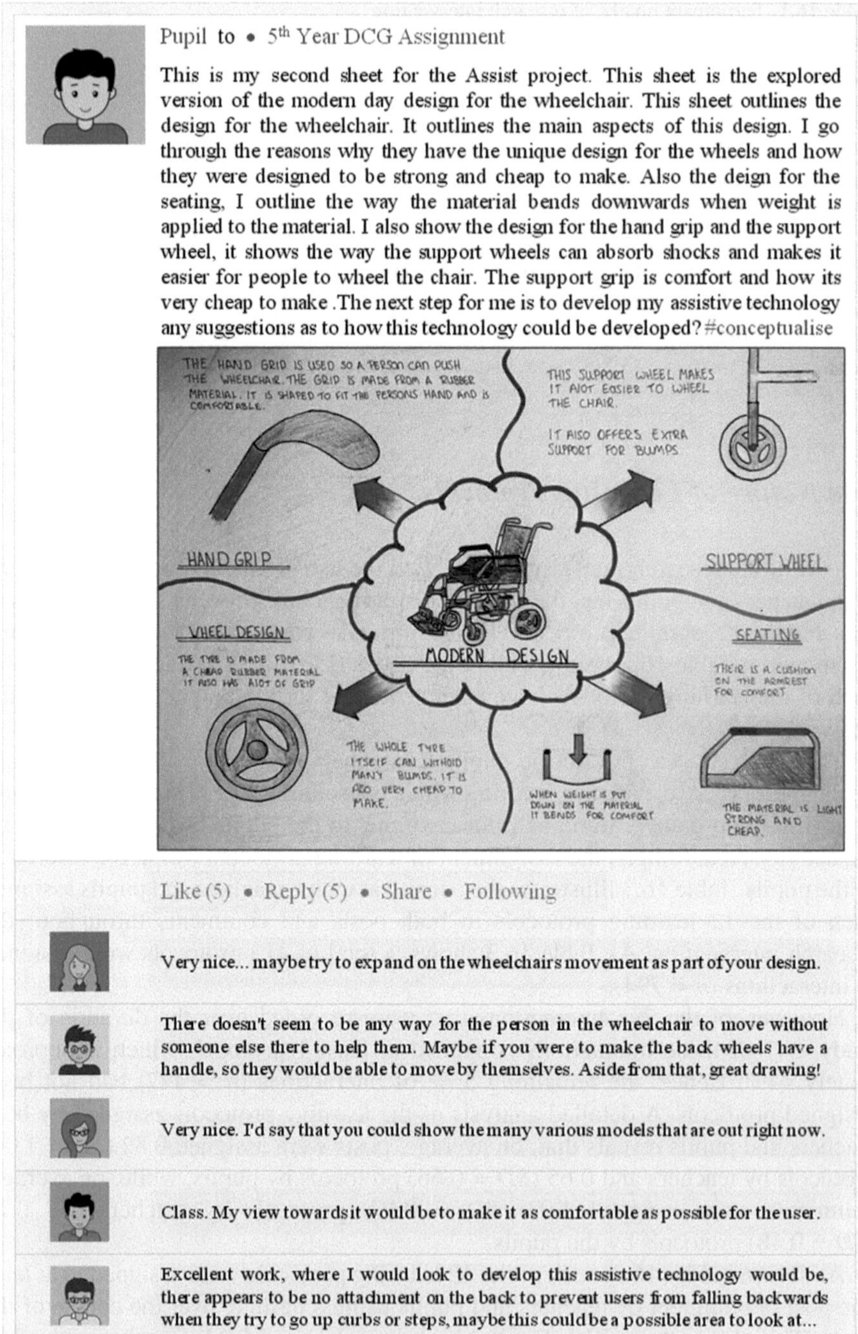

Pupil to • 5th Year DCG Assignment

This is my second sheet for the Assist project. This sheet is the explored version of the modern day design for the wheelchair. This sheet outlines the design for the wheelchair. It outlines the main aspects of this design. I go through the reasons why they have the unique design for the wheels and how they were designed to be strong and cheap to make. Also the deign for the seating, I outline the way the material bends downwards when weight is applied to the material. I also show the design for the hand grip and the support wheel, it shows the way the support wheels can absorb shocks and makes it easier for people to wheel the chair. The support grip is comfort and how its very cheap to make .The next step for me is to develop my assistive technology any suggestions as to how this technology could be developed? #conceptualise

Like (5) • Reply (5) • Share • Following

Very nice... maybe try to expand on the wheelchairs movement as part of your design.

There doesn't seem to be any way for the person in the wheelchair to move without someone else there to help them. Maybe if you were to make the back wheels have a handle, so they would be able to move by themselves. Aside from that, great drawing!

Very nice. I'd say that you could show the many various models that are out right now.

Class. My view towards it would be to make it as comfortable as possible for the user.

Excellent work, where I would look to develop this assistive technology would be, there appears to be no attachment on the back to prevent users from falling backwards when they try to go up curbs or steps, maybe this could be a possible area to look at...

Fig. 16.2 Example of post with pupil and teacher comments

Table 16.2 Codes for learning protocols

Code	Protocol	Description
LP01	Conceptualise	Creating a mental grasp of something to develop an idea or feeling
LP02	Externalise	Giving abstract meaning to some or all elements of an idea or feeling
LP03	Formalise	Applying tangible substance to practise or test an idea or feeling
LP04	Analyse	Examining an idea or feeling in order to explain and interpret it
LP05	Synthesise	Making what is known about an idea or feeling into a coherent whole
LP06	Rationalise	Qualifying the importance or significance of a particular idea or feeling
LP07	Generalise	Describing broad applications or conclusions from an idea or feeling
LP08	Hypothesise	Suggesting a possible explanation or direction for an idea or feeling
LP09	Theorise	Reasoning supporting principles to substantiate an idea or feeling
LP10	Revise	Looking over an idea or feeling in part or in full to correct or better it
LP11	Reorganise	Changing the way in which an idea or feeling has been formulated
LP12	Recognise	Knowing how to progress or reconstruct an idea or feeling

Table 16.3 Density of learning protocols

	Teachers			Pupils			
School	*N*	Post	Comment	*n*	Post	Comment	Total
S_01	1	14	44	13	29	15	102
S_02	1	1	11	24	56	27	95
S_03	1	2	4	18	34	2	42
S_04	1	5	1	7	36	28	70
S_05	1	3	11	5	20	0	34
S_06	1	5	0	9	13	14	32
S_07	1	11	19	11	8	0	38
Total	7	41	90	87	196	86	413
A.M.	1.00	5.86	12.86	12.43	28.00	12.29	59.00
S. D.	0.00	4.85	15.27	6.63	16.20	12.12	29.86

and pupils' use of protocols that shows evidence of an iterative and interactive learning process. As is apparent when looking at LP10–LP12, the protocols assigned by pupils are the evidence of their learning process in response to the protocols assigned by teachers.

Agreement of Learning Protocols

In comparing the frequency of protocols assigned by participants and the researcher, Fig. 16.4 indicates there was a clear sense of hierarchical progression in the protocols.

And, in comparing the agreement of protocols assigned by pupils and the researcher, Fig. 16.5 indicates that pupils' understanding of this 'meta-language' of progression increased and began to align with the researchers' understanding of learning protocols.

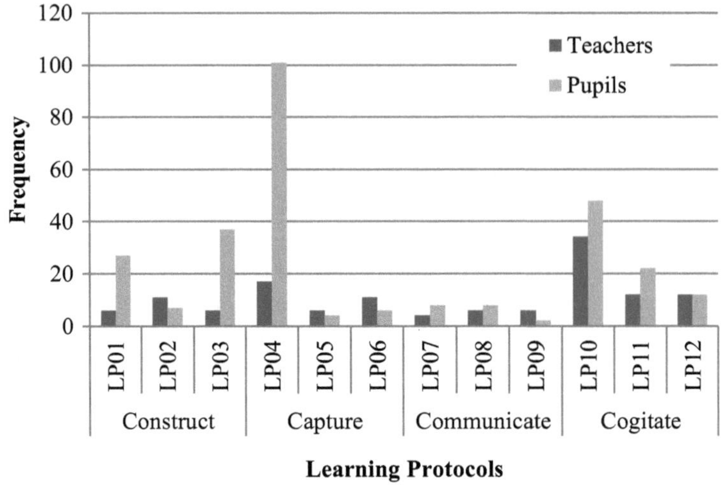

Fig. 16.3 Frequency of protocols assigned by teachers and pupils

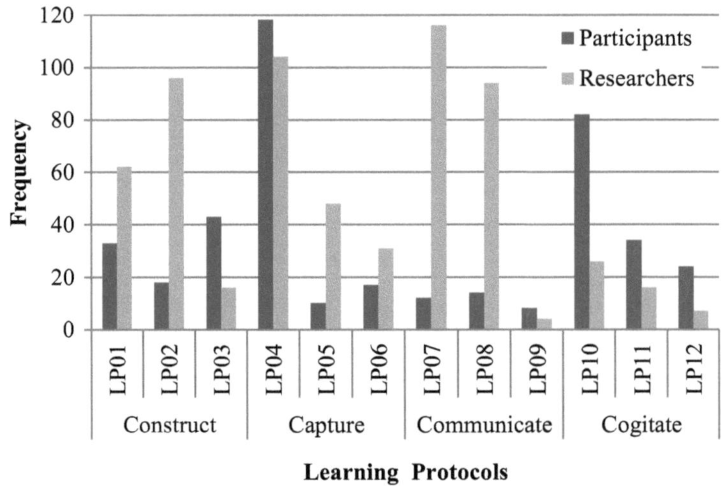

Fig. 16.4 Frequency of protocols assigned by participants and researcher

How This Might Be Used to Improve Teaching and Learning

In general, traditional practice allows for education and instruction to be delivered in a physical classroom setting, and the current practice of online and/or blended learning allows for education and instruction to be delivered primarily via the Internet in a virtual classroom setting and/or delivered in part via the Internet in a

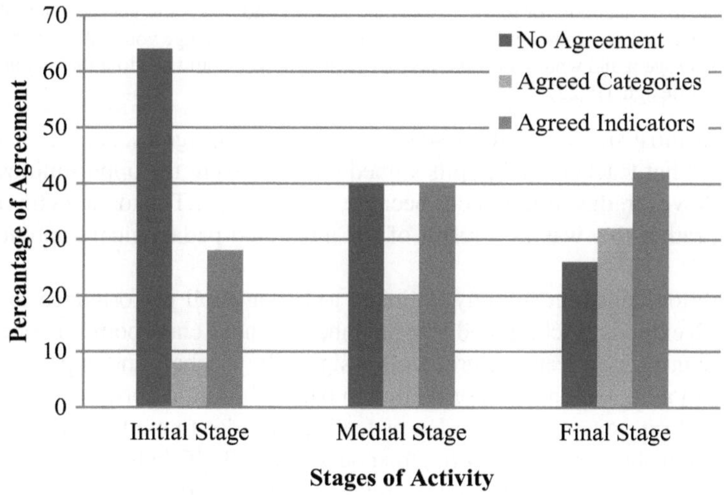

Fig. 16.5 Agreement of protocols between pupils and researcher

virtual classroom setting and/or delivered in part in a physical classroom setting (O'Connor et al. 2016). However, with the latest advancements in ICT and cloud-based software, education and instruction are no longer confined to a 'classroom' setting as new configurations for the delivery of education and instruction are possible in any conceivable setting. This enables the development of a new space or dimension for teaching and learning where borders are only limited by the imagination of those who participate in them, blurring the traditional institutional, spatial, and temporal boundaries of schooling.

The Effects on Teaching and Learning

Considering the frequency of posts and comments recorded during this intervention, this research shows that teachers and pupils responded positively to the integration of new technological and pedagogical approaches within the traditional environment as well as the ability to extend the teaching and learning beyond the classroom setting. Teachers and pupils welcomed this contemporary approach, indicating this stimulated their interest, motivation and engagement in the regular day-to-day school activity. For example, as was presented in the following extracts taken from T_02 and P_22:

> … I really enjoyed it and I found it very interesting. I think it's great for pupils and for teachers as it's hard to give everyone the feedback they need in class, but when it's online you can give everyone a bit more feedback than usually, and you can identify the pupils that need your help a lot more easily. (T_02)

> … it was like Facebook for school, you could interact and you could get feedback off your friends and then you can begin to change your design and challenge yourself to see how you could improve it and what you could take away or what you could add to it that would make it a better design. (P_22)

What is most significant in these extracts is the language used, as it clearly indicates that what teachers and pupils valued the most were the opportunities for discourse. However, this didn't occur because systems of ICT made it technologically possible, rather this was as a result of the integrated pedagogical approach of the EPI Model.

In addition, since technology is linked to better pupil performance in contexts where ICT extends teaching and learning and enhances classroom practice (OECD 2015), the most used learning protocols assigned by teachers and pupils were LP10 (Revise), looking over an idea or feeling in part or in full to correct or better it, and LP04 (Analyse), examining an idea or feeling in order to explain and interpret it. This research shows that supporting discourse using TMC helps to extend learning and enhance practice as it promotes greater periods of critical thinking and reflection and, hence, attends to the integration of ICT and the effective pedagogical use of ICT. Furthermore, as was presented in the following extract taken from P_78 and T_05, pupils indicated that learning protocols made the process of learning more effective:

> … it was really good for learning because you can search for the hashtag, to see all the posts with that hashtag and find what you were looking for. (P_78)

> … they were thinking more, because they were allowed more time. If you're given more time to answer a question you're going to answer it better. (T_05)

Subsequently, the following extract presented below taken from T_06 (post-study) indicates how the asynchronous nature of the approach enhances classroom practice:

> … I can have the homework checked before coming into school, being able to contact them at home and solve the problem at home, you're able to move on, and you're constantly going forward with new material in every class. (T_06)

For these reasons, the integration of the EPI Model allows ICT to be fully realised and exploited for the purpose of making real contributions to both teaching and learning and can ensure our schools don't lag considerably behind the promise of technology. Although teachers and pupils welcomed the use of ICT in their classroom activity, it was recognised that it functioned as the medium for the activity only, to support the richness, involvedness, and connectedness of the pedagogical approach and provide the right support and the right feedback, in the right time and place, at the right level. This was the priority concern of the research intervention from the outset, to ensure that any form of technology introduced within the classroom activity should operate as an enhancement for the activity only, rather than as a distraction or as a distortion. The integration of ICT did not act as a barrier for supporting discourse in this study, as teachers were seen to fully control the impediments associated with using TMC.

The Process of Teaching and Learning

This study implemented the use of learning protocols (i.e. hashtags) to translate the EPI Model into a pedagogical and technological approach to help teachers and pupils articulate a common language of progression around the intended learning processes. Accordingly, this research shows that by creating a shared language of understanding, the relationship between the frequency of teachers' and pupils' use of these protocols, as shown in Fig. 16.3, is evidence of an iterative and interactive process of learning.

In comparing the frequency of protocols assigned by participants and the researcher, a clear sense of hierarchical progression in the protocols was shown in Fig. 16.4, which further enabled participants to evidence their thinking in terms of progression. The classification of this hierarchy indicates that effective use of the learning protocols along with engaged teachers and supported discussions has been found to be the most consistent indictors of pupils reaching higher phases of critical thinking and reflection. Although the spatial and temporal independencies provided by the integration of ICT gave pupils more time to critically think and reflect on their interactions with others, by working iteratively, using the EPI Model to assign learning protocols to posts and comments, working with others, and uploading subsequent data files as evidence encouraged pupils to critically think and reflect on their interactions with themselves. This included having to share the evidence of your learning within the public sphere and having to assign a learning protocol that was appropriate to the evidence shared. For example, as presented in the following extract as taken from P_77 (focus group), the pupils identified two common factors that promoted metacognitive awareness:

> … with the hashtags, we were analysing our own work, and we were sharing it with everyone else, we were communicating our ideas with everyone else, but we were also communicating with ourselves, it's awareness for ourselves as well at the same time, it was like killing two birds with the one stone. (P_77)

In comparing the agreement of protocols assigned by the pupils and the researcher, Fig. 16.5 revealed that pupils' understanding of this meta-language of progression increased and began to align with the researchers' understanding of learning protocols. The alignment between the pupils' and researchers' use of learning protocols begins to offer some validation of the usefulness of the EPI Model for supporting discourse. This is significant because giving pupils a language they can use gives them a voice.

Furthermore, the following extract presented below taken from P_17 (focus group) shows pupils could talk about protocols using the language given by the descriptors:

> … I understood them in the end, if someone wanted you to examine what is right or what is wrong in their post, they would use hashtag analyse. (P_17)

The integration of the EPI Model for supporting discourse not only generated a common language for a shared understanding of progression, but it provided a medium for pupils to have their voices heard in the process of learning. The results of a 3-year study (Nystrand 1996) that focused on 2400 pupils in 60 different classrooms indicated that the typical classroom teacher spends under 3 min an hour allowing pupils to talk about ideas with one another and the teacher and is dependent upon teacher-centred questions which had predetermined answers. It became increasingly clear from the results of the interviews and focus groups that both teachers and pupils acknowledged that in a standard lesson period, there simply isn't enough time for pupils to discuss their ideas with one another and the teacher. However, as a result of the spatial and temporal independencies provided by the ICT, this study has indicated that pupils had more time to both ask and answer questions on the issues they found to be personally meaningful and educationally worthwhile, which ultimately guided pupils in the development of more robust ideas and solutions. In addition to not participating within much discourse with others in the classroom, pupils also discussed the worrying anxieties and performance issues they usually experience when being asked a question in front of others in a traditional classroom. However, because the focus of the learning protocols, and therefore the EPI Model, was on the process of learning and not on the 'right answer', pupils felt that it was more encouraging to present and discuss their ideas as there was no 'wrong answer'. The results of the pupil surveys and findings of the focus groups both indicate that pupils felt comfortable interacting online because it was a familiar approach to TMC. For example, this is presented in the following extract taken from P_32 (focus group):

> … if you are told to stand up in class to give out like, what your idea is you'd be a bit nervous but online like, you're not really as worried. […] It's easier really like, you weren't worried, you could think about what idea you wanted to post up and like, even if it was a bad idea you post up like, in class maybe you wouldn't get good feedback but like online everyone gave good feedback to you and if there was something wrong they could help you with it. (P_32)

The Visibility of Teaching and Learning

As discussed so far, the impact of the EPI Model for supporting discourse using TMC created a new space where education and instruction can be delivered traditionally and via the Internet through embedding virtual presence into the physical classroom. Translating the EPI Model to facilitate the pedagogical and technological approach by using protocols generated a shared understanding of progression about the process of learning and supported teachers and pupils in their individual and collaborative work. Finally, the use of these protocols provided opportunities for pupils to evidence their learning and to make their learning more visible which supported teachers and pupils alike in planning their intervention strategies and basing their decisions on evidence. As is presented in the following extract taken from T_01, this was possible because the use of protocols allowed the reader to see and understand the process of learning and in turn informed a person's capacity to identify current and targeted progressions and whether or not they should or shouldn't intervene within the process of learning:

> … when I can see what they hashtag, I can see what the pupils are thinking, so if its #conceptualise maybe they are looking to show what their ideas are, it gives teachers and whoever else is looking at it an idea of what the pupil or the other person's actually thinking and how they're actually doing it. (T_01)

Accordingly, teachers were able to use this evidence in their planning for learning and to provide individual, developmental, and personalised learning for each pupil. As discussed in the extract below taken from T_01 (post-study) and was presented in the findings, the visibility and authenticity of learning provided by the EPI Model in conjunction with the asynchronous capacity of TMC have evolved teachers' pedagogy into a more diagnostic and formative approach to planning for teaching and learning:

> … it made my life ten times easier, it made planning so much easier because instead of trying to come into class with the assumption of what they knew, I could come into class knowing what they knew and knowing where I needed to develop. […] I would come in with the resources that benefited what I was going to do in class and nine times out of ten, what I thought they knew or what the posts showed me they knew, they did know, and it also showed me what they didn't know. (T_01)

Therefore, since half of all teaching interventions focused on giving direct instruction, participating teachers used the EPI Model to identify current and targeted progressions which was fundamental to when they decided to intervene in the process of learning.

What Else Would Be Good to Know, and How Could Teachers Find Out?

In addition to the design of the learning environment, the delivery of an educational transaction situated in the context of socially mediated activity and focused on the cognitive interplay of teachers and pupils was central to the pedagogical approach. Therefore, educational transactions were designed to facilitate design-based activity due to its strong affiliation with the development of social and cognitive competencies. As shown in Fig. 16.6, there were two distinct parts to each design assignment. Part (a) deals with the investigation, evaluation, and representation of the existing object(s). Part (b) deals with the process of design, either modifying existing design features of the object(s) in question or developing a completely new design solution to the brief.

Subsequently, as presented in Table 16.4, a standards-referenced measure of social and cognitive problem solving was used to assess individual and collaborative evidence of the tasks performed or competencies displayed throughout the educational transaction.

Assistive Technology is defined as a piece or item of equipment which is used to help an individual perform some task within their daily life. These devices are often mechanical aids, which substitute or enhance the functional capacity of some physical, visual or mental ability that has been impaired.

A. Complete a personal design investigation on any existing form of assistive technology. Your design investigation should include a brief exploration of such technology over time.

and

B. Design and graphically communicate a new method of assistive technology which is based on a selected functional capacity or advancement and aimed at a particular target market.

Fig. 16.6 Example of student assignment

Table 16.4 Progression of social and cognitive development

Stage	Social development	Cognitive development
1	Limited interaction	Exploration
2	Supported working	Establishing information
3	Awareness of partnership	Sharing and connecting information
4	Mutual commitment	Strategic planning and executing
5	Appreciated and valued partnership	Efficient working
6	Cooperation and shared goals	Refined application and problem solving

Using these stages of social and cognitive development empirically derived from the ATC21s Project (Griffin et al. 2013), this research was able to evaluate the individual and collaborative performances of pupils based on the posts, comments, and data files uploaded to the VLE. This was carried out after each transaction had been completed.

References

Brown, A. L. (1992). Design experiments: Theoretical and methodological challenges in creating complex interventions in classroom settings. *The Journal of the Learning Sciences, 2*(2), 141–178.

Ceertz, C. (1973). *The interpretation of cultures*. New York: Basic Books.

Cohen, L., Manion, L., & Morrison, K. (2011). *Research methods in education* (7th ed.). Milton Park: Routledge.

DES. (2007). *Sé Sí: Gender in Irish education*. Dublin: The Stationary Office.

Griffin, P., Woods, K., Mountain, R., & Scoular, C. (2013). *Module 1: Using a developmental model to assess student learning*. Melbourne: Assessment Research Centre.

Hamilton, J. W. (2003). Interaction, dialogue and a creative spirit of inquiry. In E. W. L. Norman & D. Spendlove (Eds.), *Design matters: Design and technology association international research conference 2003, University of Warwick, England, 3–5 July 2003* (pp. 35–44). Wellesbourne: Design and Technology Association.

Hamilton, J. W. (2004). Enhancing learning through dialogue and reasoning within collaborative problem solving. In E. W. L. Norman, D. Spendlove, P. Grover, & A. Mitchell (Eds.), *Creativity and innovation: Design and technology association international research conference 2004, Sheffield Hallam University, England, 4–6 July 2004* (pp. 89–101). Wellesbourne: Design and Technology Association.

Hennessy, S., & Murphy, P. (1999). The potential for collaborative problem solving in design and technology. *International Journal of Technology and Design Education, 9*(1), 1–36.

Lei, J. (2009). Digital natives as preservice teachers: What technology preparation is needed? *Journal of Computing in Teacher Education, 25*(3), 87–97.

McCormick, R. (2004). Collaboration: The challenge of ICT. *International Journal of Technology and Design Education, 14*(2), 159–176.

Murphy, P., & Hennessy, S. (2001). Realising the potential – And lost opportunities – For peer collaboration in D&T setting. *International Journal of Technology and Design Education, 11*(3), 203–237.

Nystrand, M. (1996). *Opening dialogue: Understanding the dynamics of language and learning in the English classroom.* New York: Teachers College Press.

O'Connor, A., Seery, N., & Canty, D. (2016). The experiential domain: Developing a model for enhancing practice in D&T education. *International Journal of Technology and Design Education*, 1–15.

O'Connor, A., Seery, N., & Canty, D. (2018). The procedural domain: Developing a model for enhancing practice in D&T education. *International Journal of Technology and Design Education* (submitted for publication).

OECD. (2015). *Students, computers and learning: Making the connection.* Paris: OECD Publishing.

Teo, T. (2009). Modelling technology acceptance in education: A study of pre-service teachers. *Computers & Education, 52*(2), 302–312.

Trebell, D. (2007). A literature review in search of an appropriate theoretical perspective to frame a study of designerly activity in secondary design and technology. In I. E. W. L. Norman & D. Spendlove (Eds.), *Linking learning: Design and technology association international research conference 2007, University of Wolverhampton, England, 4–6 July 2007* (pp. 91–94). Wellesbourne: Design and Technology Association.

Zhao, Y., Pugh, K., Sheldon, S., & Byers, J. (2002). Conditions for classroom technology innovations. *The Teachers College Record, 104*(3), 482–515.

Chapter 17
The Impact of Mobile Devices on Self-Directed Learning and Achievement

Scott R. Bartholomew

Abstract As mobile devices have become increasingly ubiquitous in society, the potential for impacting student learning in classroom settings has been the subject of debate. Instant information access and collaboration have been mentioned as potential benefits of mobile device access. However, the potential for mobile devices to foster self-directed learning in students and improve student achievement on open-ended design problems has not been explored. Therefore, this chapter presents the findings from research with middle school students working in small groups on an engineering design challenge. The findings explore the relationships between access to mobile devices, student achievement, and student self-directed learning.

Mobile Devices and Self-Directed Learning

Today's K-12 students are growing up in a world connected through technology. The computing power of mobile devices, carried in the pockets of many of today's students, represents more computational capabilities than their parents and teachers ever experienced during their adolescent years. Access to mobile devices among students continues to rise—Pew's research (2015, p. 1) found that 73% of American teens have access to a smartphone, "92% of teens go online daily, and 24% say they are online 'almost constantly.'" This trend appears to be a worldwide phenomenon as mobile devices become increasingly accessible (Liu et al. 2014).

As with any technological advancement, there has been discussion both for and against the inclusion of mobile devices in classrooms (O'Bannon and Thomas 2015; Elder 2009; Johnson et al. 2011; Lloyd 2010; Quillen 2010; Schenker 2009; Shuler 2009). Despite the range of opinions and findings related to mobile devices, the relationship between access to mobile devices and student self-directed learning has not been fully explored.

S. R. Bartholomew (✉)
Purdue University, West Lafayette, IN, USA
e-mail: sbartho@purdue.edu

P J. Williams, D. Barlex (eds.), *Explorations in Technology Education Research*, Contemporary Issues in Technology Education,
https://doi.org/10.1007/978-981-13-3010-0_17

Self-directed learning (SDL), an important twenty-first century trait (Prensky 2007), is a process where individuals take the initiative to diagnose their own learning needs, identify resources for learning, carry out their learning, and then evaluate their own learning (Knowles 1975). SDL is becoming increasingly relevant in today's educational landscape (Mitchell 2014) as students will graduate and encounter occupations and opportunities that don't currently exist. Importantly, a recent study by Fahnoe and Mishra (2013) noted that technology, represented by mobile devices or other items, may play a role in assisting students to be more self-directed in their learning.

This study was situated in middle school technology and engineering education (TEE) classrooms in the USA where students worked in small groups of 2–3 during an open-ended engineering design activity. The specific questions guiding this study were centered on the relationship of access to mobile devices (e.g., iPad, smartphone) and student self-directed learning as well as student achievement (i.e., score received). The two research questions that guided this study were:

1. What is the impact on student self-directed learning, if students are provided access to mobile devices?
2. What is the impact on student achievement, if students are provided access to mobile devices?

The findings from this research are important as school administrators, teachers, parents, and students debate the place of mobile devices in the classroom. On one hand the purpose of this study was to inform school and district policies regarding mobile devices. On the other hand, this study was meant to be a specific example for students and teachers to use as they continue to determine how mobile devices will, and will not, be used by students in the classroom.

Although this study specifically investigated mobile devices, the findings should not be strictly confined to mobile devices. Mobile devices represent one specific type of technology—one which provides access to communication and information. The findings related here can, and should, be used to inform and guide a variety of topics around technology, information, and access in classrooms.

How the Research Questions Were Answered

This study took place in a large suburban school district with approximately 700 middle school students at five schools (18 classes, 6 teachers). The study took place over five class periods (2 weeks of every other day class schedule, 90-min class periods) and revolved around an open-ended design problem. Two teachers used paper portfolios with their classes, while four teachers used iPads to complete identical electronic portfolios. Four randomly assigned teachers (i.e., one paper-based portfolio, three iPad-based portfolios) allowed free mobile device access to their students, while the other two teachers (one paper-based portfolio, one iPad-based portfolio) continued with the traditional rules against mobile device access during the unit.

Students were placed into groups of 2–3 and began the study by completing a series of pre-study questionnaires. Students received instruction in their groups related to mobile device use, digital citizenship, and the engineering design process, and then students were provided with the challenge for this project. This design challenge involved several criteria/constraints and centered on designing a new container/dispenser for distributing pills to an elderly individual in specified quantities and at prescribed times (see Fig. 17.1).

Students began brainstorming and compiling their portfolio and were provided with materials for constructing a solution to the design problem. The student portfolios, both paper-based and electronic, were filled out by the student groups during the designing process and followed the same structure with identical prompts and spaces for recording information. On the fifth-class period, students completed their portfolios, design products, and a post-study questionnaire. Five students from each teacher were interviewed, in a semi-structured interview format, with questions related to self-directed learning, mobile devices, engineering design, and their experience with the study. Teachers were also interviewed and asked similar questions—these questions centered on what self-directed learning "looks like," open-ended design problems, and mobile devices. All responses were transcribed and coded using descriptive and thematic coding techniques (Saldaña 2013). The resulting themes were compared with the quantitative findings to further understand the research results.

All the student products were collected, and a digital picture was taken of each one, resulting in 175 images of student design products (see Fig. 17.2). Each paper portfolio was also "digitized" into electronic portfolios for uniformity in grading (see Fig. 17.3). Student and teacher responses to the interviews were transcribed and analyzed for emerging themes. The resulting themes from the interviews were used to further expand on the findings from the study. Representative phrases and remarks from the student and teacher responses were included as examples of key findings.

A panel of five graders with a design background used adaptive comparative judgment (ACJ) to assess all the student portfolios and products. The ACJ process involved graders making a series of judgments between pairs of items. For each pair of items, judges chose the "better" item and provided comments regarding their decision. The result from the ACJ process was a rank order of all student products and a rank order of all student portfolios. The resulting rank orders from the panel of judges were obtained with a reliability coefficient of $r = .959$ for student products and $r = .972$ for student portfolios. This rank order was added to the statistical data set for later analysis.

The Design Challenge

Context: An elderly individual enjoys traveling internationally. Ideally, this person would like to travel internationally between 2-3 months of the year. This person has a few ailments and allergies that require medication. In addition this person also takes vitamins.

Challenge: You have been hired to design a new medicine dispenser for this client. Your design should:

1. Be easy to use
   a. Easy to open and close
   b. Easy to get pills in and out
2. Assist this person in remembering when to take the pills
   a. Day of the week and time of day
   b. Correct number of pills that should be taken.

Criteria & Constraints: Your design should:

1. Remind the person when to take each pill (that is: time of day and day of the week).
2. Remind the person how many of each pill to take.
3. Be small enough to fit easily in a purse, handbag, backpack, or pocket for travel (should fit easily within an 8" cube)
4. Be childproof (that is: difficult for a child to open).

Resources: The breakdown for when pills should be taken and the quantities is included here.

Pill Name	Pill Size	Number taken at each dose	When to take the pill
Vitamin A	0	2	Monday (morning)
Vitamin B	2	1	T/TH (night)
Vitamin C	1	1	Sunday (morning)
Iron	2	1	M/W/F (morning)
Allegra D	0	1	Daily (morning)
Potassium	1	1	Daily (night)
Sodium	0	1	T/TH (morning)

	Sunday	Monday	Tuesday	Wednesday	Thursday	Friday	Saturday
AM	Allegra D Vitamin C	Allegra D Vitamin A Iron	Allegra D Sodium	Allegra D Iron	Allegra D Sodium	Allegra D Iron	Allegra D
PM	Potassium	Potassium	Potassium Vitamin B	Potassium	Potassium Vitamin B	Potassium	Potassium

Pill Size 0	Pill Size 1	Pill Size 2
M&M Mini	M&M Candy	M&M Peanut Butter
Height: .35"	Height: .47"	Height: .6"
Width: .35"	Width: .47"	Width: .6"
Thickness: .2"	Thickness: .25"	Thickness: .3"

- *For this design challenge you can assume that all pills are the sizes and shapes shown above and listed in the table*

Supplies:

Students will be provided with tools, materials, and supplies to proto-type and build while they are designing. Students should plan carefully to conserve materials as no additional materials will be provided. All material does not need to be used in the design. Building items include:

General Supplies

- Plastic bag containing all supplies
- 10 3x5 cards
- 2 copies of the engineering design process
- 2 copies of the engineering design challenge
- 1 pair of dice
- 2 red pencils
- 2 green pencils
- 1 Pentax Fujifilm instant camera (paper groups)
- Film (paper groups – 130 sheets per teacher)
- 1 pad of post-it notes
- 1 plastic cup
- Plastic (one 12" x 12" sheet - .007" thickness)
- Cardstock (two 8.5" x 11" sheets, assorted colors)
- Rubber bands (approximately 25, assorted sizes/shapes)
- String (polyester kite string, 3')
- Paper clips (20 small, 10 large)
- Straws (ten flexible neck)
- Dowel (four .125 X 4")
- 20 m&m's minis (to represent pill size 0)
- 15 m&m's (to represent pill size 1)
- 10 m&m's peanut butter (to represent pill size 2)
- 5 buttons
- 4 clothespins
- 20 jumbo craft sticks
- 15 toothpicks
- 10 small cups with lids
- 10 interlocking craft sticks
- 10 Pipe cleaners
- Tape (masking tape, 1 roll)
- Hot glue gun and glue (10 glue sticks)
- Scissors (1 pair)
- Paper (8.5" x 11" sheets, white)

Evaluation Rubric: students will complete a design portfolio that will document their process as they design their product. Students will be rated based on their design portfolio and their final product using the rubrics below.

Portfolio Evaluation

Item	Evaluation Criteria	Item Weight Value
Questions/Prompts	Each question or prompt was responded to by the students with an explanation, picture, or drawing.	2
Pictures	Each picture box contains a picture representing student work. Pictures demonstrate a logical progression of the product through the design process.	1
Design Process	Steps of the engineering design process are clearly demonstrated by the students in the portfolio. 1. Identify the need or problem 2. Research the need or problem 3. Develop possible solutions 4. Select the best possible solution 5. Construct a prototype 6. Test and evaluate the solution 7. Communicate the solution 8. Redesign 9. Finalize the design	1
Overall Portfolio	Portfolio is easy to read, follow, and understand	1
Self-directed Learning	Student demonstrated self-directed learning in their portfolio creation	1

Product Design Evaluation

Item	Description	Item Weight Value
Criteria and Constraints	Designed product satisfies provided criteria and constraints	1.5
Feasible & Functional	Designed product is both feasible and functional	1.5
Aesthetics	Design product is aesthetically pleasing	1
Creativity	Designed product demonstrates original thought, insight, and innovation	1

Fig. 17.1 Student design challenge

Fig. 17.2 Example of student product

What I Found Out

Student Self-Directed Learning Student self-directed learning was assessed prior to, and immediately following, the research using a modified version of the *Self-directed Learning with Technology* Scale (Teo et al. 2010). The statistical analysis of the research data revealed two significant predictors of student self-directed learning scores: average mobile device skill level and computer access and use at school. Average mobile device skill level, as derived from student answers to a modified version of the *Digital Natives Assessment Scale* (Teo 2013), was positively correlated with student self-directed learning ($r = .40$), suggesting that students with a higher skill level in using mobile devices were more self-directed in their learning. Computer access and use at school was negatively correlated, albeit weakly, with student self-directed learning ($r = -.18$), indicating that students who used computers, at home and at school, were less self-directed in their learning.

Additional tests were used to identify whether students were more self-directed following the research study which provided them an opportunity to work on an open-ended problem with some students permitted access to mobile devices. The paired-samples t-test showed a significant increase in student self-directedness in learning following participation in the study ($t = 6.521$, $p < .001$, $d = -.44$) which suggests that students were more self-directed after the study than prior to the study taking place. This hints that participation in open-ended problems may lead to an increase in student self-directed learning.

As different portfolio mediums were used (paper vs. e-portfolio) during the study, a statistical analysis of covariance was utilized to assess whether student self-directedness could be attributed to the medium of portfolio assigned to them. The

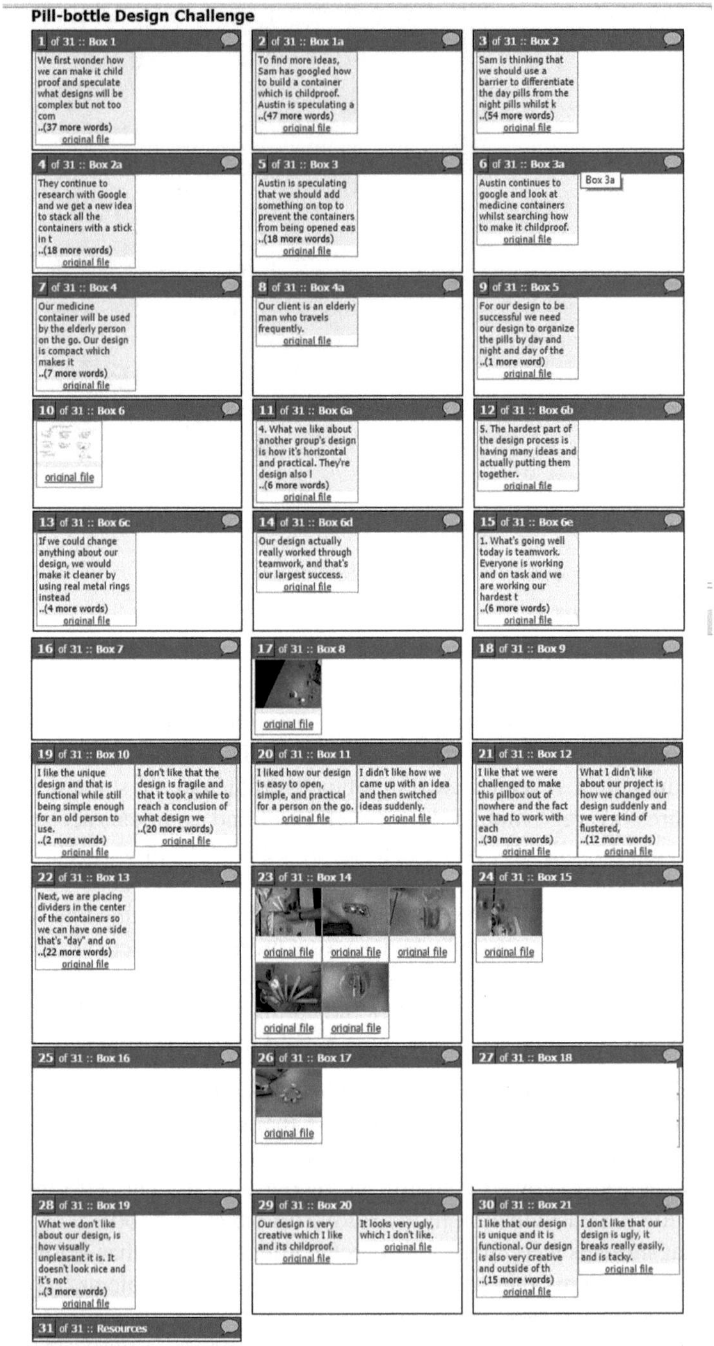

Fig. 17.3 Example of student portfolio

resulting *p*-value was not statistically significant ($p = .132$) suggesting that student self-directedness in learning was not a function of their assigned portfolio creation medium.

Additionally, ANCOVA techniques were used to assess whether student access to mobile devices was significantly related to student self-directedness. The resulting values were not significant, $F(1, 218) = .054$, $p = .816$, suggesting that student self-directedness in learning within this project was not dependent on access to mobile devices.

Using a simple bivariate correlation, the relationship between student comfort level with open-ended design problems and self-directedness was analyzed. This relationship was significant, $r = .34$, $p < .001$, in the positive direction which suggests that higher comfort levels in students with open-ended design problems were correlated with more self-directedness in learning. Similarly, the relationship between student comfort level in working with groups and self-directedness was significant, $r = .19$, $p < .001$, and positive, suggesting that comfort in group work also corresponded with self-directedness.

Student Achievement Student achievement was measured in two ways as part of this study: (1) student group rank score on their design portfolio and (2) student group rank score on their group product (created during the design challenge). The two rank positions for each group of students were used in the statistical analyses to explore possible relationships with other variables in the study.

The student *portfolio* rank was significantly correlated with several student characteristics (see Table 17.1). A significant correlation indicates that as students ranked better on their portfolio, they also increased significantly with the other variable identified (i.e., high grades in class). However, despite multiple significant correlations between the portfolio and other variables, the only significant relationship between these variables and the student *product* rank was age ($p = .05$), which showed that *younger* students were correlated with better ranking products.

Correlational analyses revealed that the relationship between student self-directedness in learning and their final portfolio rank was not significant; students that were more self-directed did not rank better than their peers on the portfolios. Similarly,

Table 17.1 Student demographics measures and student portfolio rank score

Student portfolio rank	Pearson correlation	Sig. (2-tailed)	*N*
Student age	.16	.02[a]	221
Grades in average (all classes)	.13	.05[a]	220
Grades on average (TEE only)	−.02	.83	221
Average time using technology	.27	.00[a]	214
Average mobile device use	.05	.45	221
Skill level with mobile devices	.15	.02[a]	221
Mobile device access (home) and use	.27	.00[a]	219
Mobile device access (school) and use	.24	.00[a]	219

[a]Indicates a significant relationship

the correlation between student self-directedness in learning and their product rank was also not significant; students that were more self-directed did not rank better than their less self-directed peers on the products. These correlations were not significant for students' self-directedness in learning either before or after the research study.

The influence of portfolio medium was analyzed with respect to the student achievement scores (using both portfolio and product rank scores). The results revealed that the portfolio medium was significant for both the product rank score (t (455) = −4.83, $p < .001$) and the portfolio rank score (t (454) = −5.84, $p < .001$). It is important to note that while the resulting t-values are negative, this is logical as a lower rank score represents a higher quality item - as determined through the rank order. These tests reveal that students who used a paper portfolio received significantly better ranks than their peers using the electronic portfolio.

An independent samples t-test was used to investigate the impact of access to mobile devices on student achievement (using both portfolio and product rank scores). The tests revealed that access to mobile devices had a significant impact on student portfolio rank (t (454) = −3.62, $p < .001$); students with access to mobile devices ranked better than their peers on the portfolios than who did not have access. However, there was not a significant difference in student product rank between those with access to mobile devices (t (455) = .816, $p = .415$).

Recognizing the influence of teachers on student achievement, a one-way ANOVA was computed to assess the impact of the teacher on student achievement results. The results were significant for both portfolios ($F = 37.70$, $p < .001$) and products ($F = 8.77$, $p < .001$), suggesting that the influence of the teacher was significant in how students' outputs (portfolios and products) were ranked relative to their peers at the conclusion. This finding is further strengthened as none of the teachers participating in the study were involved with the ACJ process for ranking; this suggests that even unconnected judges identified significant differences in quality of student work based on which teacher the student groups had. Post hoc analyses were computed to further explore the differences between how student work ranked based on their teacher. Table 17.2 shows the differences for portfolios by teacher and Table 17.3 shows the differences for products by teacher. Teacher 1 and Teacher 6's students were significantly better ranked than the other teachers for the portfolios. Teacher 1, 4, and 6's students were significantly better ranked than the other teachers for the products.

Interview Findings The final themes emerging from the student and teacher interviews were used to further explore the research questions. The themes, along with representative responses, are included here.

Self-Directed Learning Student and teacher comments related to self-directed learning revolved around the necessity of *student choice* for self-directed learning to occur. For example, one student commented:

> [self-directed learning is] somebody actually choosing what they have to do and what they want to do in their education.

Table 17.2 Post hoc analysis of differences in student portfolio rank by teacher

Teacher (*n, m, sd*)	1	2	3	4	5	6
Teacher 1 (84, 64.26, 48.98)		.00	.00	.00	.00	.12
Teacher 2 (84, 130.55, 44.32)			.00	.00	.00	.00
Teacher 3 (69, 85.20, 45.78)				.22	.00	.00
Teacher 4 (59, 94.58, 36.23)					.08	.00
Teacher 5 (53, 108.75, 47.83)						.00
Teacher 6 (107, 86.47, 50.56)						

Table 17.3 Post hoc analysis of differences in student product rank by teacher

Teacher (*n, m, sd*)	1	2	3	4	5	6
Teacher 1 (85, 77.99, 47.39)		.00	.00	.35	.00	.79
Teacher 2 (84, 107.17, 51.11)			.94	.00	.62	.00
Teacher 3 (69, 106.54, 48.26)				.00	.69	.00
Teacher 4 (59, 70.10, 47.20)					.00	.46
Teacher 5 (53, 102.85, 44.51)						.00
Teacher 6 (107, 76.05, 54.89)						

Similarly, one teacher highlighted the need for student initiative and choice in self-directed learning:

> Self-directed learning is where a student takes their own personal initiative to take the supplies that I've provided and also the knowledge that I have provided that they need, and of their own knowledge and their own supplies—based off of rules and I guess regulations, based off of our assignments or whatever—to create a learning environment where they are benefitted.

Mobile Devices As teachers and students reflected on mobile devices, their responses revolved around three distinct themes: (1) mobile devices are *enablers* for both positive and negative behavior, (2) mobile devices are regulated by *strict rules* and monitoring, and (3) *classroom norms* deter mobile device integration.

Comments around the enabling nature of mobile devices included responses such as:

> Well it just matters on the kid pretty much. I think that [mobile devices] would help most kids, but some kids are just there to get the grade and to dink off with it and ruin the privileges. It would help them because like, they, oh I feel familiar with this—I know what to do, I know where to go.

The students interviewed identified strict rules for mobile devices in K-12 classrooms. Student mentioned different "areas" where mobile devices were "allowed" and other "areas" where mobile devices were prohibited. One student commented:

> In school [mobile devices] are allowed during class, if the teacher gives you permission, only if you are, like, working on an assignment or something. Um, they are allowed during lunch—private time—before and after school. Um, and like usually people just, like, use them to do, like, calculators or math, and stuff like that.

While teachers noted benefits of mobile devices most of their comments revolved around the need for such rules. One teacher noted:

> I think that [mobile devices in K-12 classrooms] can be good in a monitored fashion, with activities like the one we did, or other experience design activities. It could be very valuable in the research and understanding what the actual problem is they're trying to solve and where it fits in the world of what the impact that decision or solution might have.

It was noted that, regardless of permission to use mobile devices, most students did not choose to use mobile devices regularly throughout the study. Noting this the researcher asked students and teachers about this decision to use, or not use, mobile devices. One teacher speculated:

> I had a couple kids looking on the I-pad on the internet. Honestly, I was surprised that when we opened it up to the mobile devices more students didn't have their cell phones out. Most of them were just looking for images or for, in of the pill bottle folder things. But I was surprised at, I guess, the lack of using that device. Maybe it's because they're not used to using it in my classroom—I really don't know. The only thing I can think of is because it's the rule that you don't have your cell phone out in my class, I kind of felt like that was it - the norm.

Interestingly the findings, teacher comments, and researcher observations all appeared to demonstrate that mobile devices may not be an influential factor in student success or self-directedness in learning. When given the opportunity, most students did not utilize mobile devices beyond an initial search engine query into images of medicine dispensers. Whether a result of classroom norms, student choice, or other factors, mobile devices may not be a solution for improving student achievement or self-directedness—other variables such as skill level with mobile devices, experiences with open-ended problems, portfolio medium (paper vs. electronic), and classroom teacher appear to be poised to make a bigger impact on student success and self-directedness in learning.

How This Might Be Used to Improve Teaching and Learning

The findings from this study can provide insights into student self-directed learning, mobile devices in classrooms, and student work on open-ended design problems.

Recommendations for Self-Directed Learning

When interviewed teachers highlighted several classroom traits and tools which they believed would improve student SDL. However, access to mobile devices did *not* have a significant impact on student SDL and access to computers at school was *negatively* correlated with student SDL. These findings, which ran counter to expectations, revealed that student SDL may not be a factor of environment traits alone (e.g., access to technology). The research did show that several student characteristics and opportunities were significantly correlated with student SDL including student skill in using mobile devices, student comfort with technology, student comfort working in groups, and student comfort with open-ended design problems.

These findings suggest that teachers can, and should, do more than simply purchase the latest technology tools and introduce them into their classrooms; more technology does not mean more student SDL; it might mean less. To best assist students in becoming more self-directed in their learning, teachers can not only equip their classrooms with technology and other tools; they should also emphasize specific skills such as working in groups, solving open-ended problems, and using technology tools. In addition to teaching these skills, teachers must provide students with opportunities to work in these settings—situations involving group work, open-ended tasks, and access to a wide variety of tools. Teacher interviews further solidified these ideas as teacher responses suggested they believe these skills, such as working in groups and solving open-ended problems, should be taught in class and opportunities should be provided to students.

In addition to these thoughts from teachers, the student interviews highlighted several other ways self-directed learning could be fostered in students. More than anything else, students want to make choices regarding their education - specifically choices about what they will study, how they will study it, and what tools they will use. Providing students with these opportunities to identify topics, approaches, and tools for learning may strengthen student SDL abilities.

Recommendations for Mobile Devices

While mobile devices did not significantly impact student SDL in this study, mobile devices did correlate significantly with higher student achievement on the design portfolio. During student interviews a theme that emerged was that of the need for direct instruction for students regarding how, where, and when to use their mobile devices. Many students were provided access to mobile devices but didn't ever use them—they didn't know how or make the connection between access to a mobile device and a potential for using that tool for learning.

A key difference in student and teacher perceptions revolved around mobile device-specific benefits; while teachers noted that mobile devices were "not necessary" if computers were present, many students felt that mobile devices were useful in ways not included in computer access. The differences between teacher and student perceptions highlight the need for explicit dialogue, discussion, and instruction regarding how mobile devices can and should be used in classrooms. Relatedly, students mentioned "mobile-friendly" and "mobile-restricted" areas existing in their school. In observations and teacher interviews, the theme of "classroom and school norms" emerged—unspoken rules that governed the way students and teachers viewed and used technology such as a mobile device. If teachers and students want mobile devices to become useful tools for improving student SDL and achievement, work needs to be done to change the existing "norms" around where/when mobile devices can, and should, be used. It's possible that this study would have returned far different results if the existing norms around mobile devices, and their place in classrooms, were different. Conscious efforts toward changing the existing norms and teaching students to productively leverage these technology tools may lead to increased achievement and SDL for students.

When interviewed students noted different "types of problems" which mobile devices were useful for and other "types of problems" which mobile devices were not as applicable. The analysis of the responses showed that students perceived mobile devices as useful tools for solving problems with one correct answer (e.g., 2 + 2=) rather than open-ended problems which may have a multitude of potentially correct answers (design a _____ to solve this problem). In order for mobile devices to become learning tools, especially in light of student SDL, teachers should work to explicitly teach students ways that mobile devices could be leveraged to perform tasks other than simply finding facts. Possible skills teachers could emphasize include exploration, brainstorming, collaborating, creativity, and criteria and constraint identification. Teacher demonstrations of mobile devices, as a tool for SDL and open-ended problem solving, would likely be useful and assist students in seeing mobile devices as tools for more than simple single-answer problems.

Recommendations for Student Achievement

The student achievement was measured through their group portfolio and product ranks after the project. Student access to mobile devices did *not* correlate with a better ranking on the product portion of the assignment, but it *did* correlate with a better ranking on the portfolio portion of the assignment. While there are many possible reasons for this, one possible reason was derived from the student interviews: the type of problem. As noted previously students appeared to classify mobile device use as "factual" rather than "open-ended." It's possible that students perceived the portfolio, with its direct questions and specific prompts, as the type of assignment that they would use mobile devices to fulfill, while the product creation portion of the assignment, with its largely creative and flexible nature, may have been perceived as "too open-ended" for mobile devices to be used effectively by the students. These findings further suggest that mobile devices may help improve student achievement but must be accompanied by teacher instruction and training.

One key finding, supported by both the portfolio and the product rank scores, suggests that the portfolio medium (paper or electronic) was highly influential in terms of student achievement. Students completing portfolios on paper produced significantly better ranking portfolios *and* products than their counterparts using electronic portfolios. It is possible that the "physical" nature of the paper portfolio as opposed to the "digital" nature of the electronic iPad portfolio was suited better to student needs. The long-standing use of notepads, sketchbooks, and paper and pencil tools in education, artistic, and design endeavors may have impacted students in ways not associated with the digital portfolio tools. Students' comfort level and familiarity with paper, pencils, and pens were likely high, due to their presence and use in classrooms, and it is possible that the tangible nature of the portfolio and the comfort associated with these "familiar" objects was enough to positively impact students toward their use in the portfolio creation process.

It's also possible that the paper portfolios were "easier" to fill out than the electronic portfolios. An inherent aspect of the electronic portfolio was the increased time to turn the iPad on, navigate to the portfolio creation app, login, find the correct portion of the portfolio, and type in a response. Observations by the researcher also pointed out that it was more difficult for students to "draw" on the iPad than it was for their counterparts to do on paper—the lack of a "pencil" drawing tool may have been enough to discourage sketching and drawing among students assigned to the iPad portfolios. These differences in the electronic portfolio may have contributed to an overall "slower" or "more cumbersome" process that influenced their achievement.

One additional distinct advantage of the paper-based portfolio resided in its transparent nature. As teachers, students, and the researcher walked around the room, it took but a passing glance to quickly identify overall progress and completion of the paper portfolio, whereas the electronic portfolio, which only displayed one design prompt/section at a time, had a very opaque nature—effectively "hiding" student progress by only displaying one section at a time. Analysis revealed that the paper portfolios were overall significantly more complete than the electronic portfolios suggesting that it is possible that one reason paper portfolio groups ranked better than their electronic portfolio counterparts may be related to the transparent/opaque nature of the portfolio medium. Teachers must consider the advantages, and disadvantages, of the medium for portfolio creation before implementing either approach in their classroom. The amount of drawing/sketching, note-taking, and overall use should all factor into the decision around portfolio medium.

Similar to the portfolio medium, the impact of each student's teacher on their performance was highly significant. Although all teachers in this study were Level 2 teachers and had similar teaching experience, background, class assignment, and recommendations from peers and colleagues, there were significant differences in the achievement of their students. While all teachers followed the same script, observations by the researcher revealed that one teacher was noticeably better at guiding the students through the portfolio and product creation process. While several other teachers struggled at times to stay caught up with the pace of the project, this teacher never struggled with pacing and required the least assistance from the researcher. Of the 176 total portfolios and 176 products, this teacher had students that produced 6 out of the top 10 portfolios and 4 out of the top 10 products. This finding was further intensified when the age of these teachers' students is considered; this teacher had the youngest students, on average, in the study. While all students participating in the study were between the ages of 12 and 15 (seventh and eighth grade), all the students taught by this teacher were 13 or younger. Additionally, it is important to note that socioeconomic status (SES), a variable often associated with student achievement and success (Darling-Hammond 2000), was not significantly correlated with student achievement in this study. The school associated with the highest socioeconomic status did not produce students that were significantly different than others—in fact, the students from this school performed worse than many other schools included in the study. Researcher observations noted that the teacher at this school was painfully less capable than the other teachers in the study. Taken together this highlights the need for teacher training, professional development, and preparation. Rather than emphasizing tools (e.g., computers or mobile devices) for improving students SDL or assignments, collective efforts should be refocused to improving teacher pedagogy, practice, and experience as this appears poised to more effectively influence student experience and achievement.

Moving Forward

This research seems to suggest that to help students improve their own SDL, teachers should teach students specific skills around working in groups and utilizing technology to perform a variety of tasks specific to the desired outcomes of the assignment. Teachers need to understand their own students' strengths and weaknesses, and their use of mobile devices, so the already-present skills can be leveraged to improve student achievement. Allowing students more choices and then investigating the implications behind what they study, how they study it, and what tools they use to assist them in their study will all help teachers make the best choices for their own classroom.

In addition to findings around SDL, this research suggests that technology tools are simply that: tools. Understanding how students use these tools (for better or for worse) and then explicitly modeling and instructing students in best practices are both necessary steps to promote success. Further, teachers should question their students to understand their perceptions of how mobile devices might be used to assist in learning and then help students see beyond the obvious "fact-finding" capabilities to promote using mobile devices in more innovative ways for deeper learning.

Teachers should reflect not only on the content of the assignments they prepare for their students but also on the medium used for both providing the assignment to the students and the medium for receiving student submissions. This research found that the medium used for portfolio creation was significant for student success, and it may be that with advances in technology and technology tools, there are also accompanying weaknesses and shortcomings in using these for student work. Further, the findings suggest that paper and pencil, as archaic as they may appear, may have distinct advantages in terms of recording student thoughts, stimulating creativity through sketching, and synthesizing student ideas. Further, the capability for teachers to immediately assess student ideas and progress—by looking at their physical sketches and portfolios—may prove too valuable to be replaced by current technology tools.

Finally, this research showed that nothing was more impactful on students than their assigned teacher; therefore, efforts around improving SDL, integrating mobile devices, or improving student learning will likely be most impacted by the classroom teacher working with the students. A "good" teacher will make more of a "lesser" situation and compensate for other lacking areas (e.g., presence of mobile devices or other tools); the researcher observed that teachers with higher-ranking students appeared to be more organized and demonstrated better classroom management than their peers with lower-ranking students.

References

Darling-Hammond, L. (2000). Teacher quality and student achievement: A review of state policy evidence. *Educational Policy Analysis Archives, 8*(1). Retrieved from http://epaa.asu.edu/ojs/article/view/392/515.

Elder, J. (2009). *OMG! Teachers say texting can be good for teens*. Retrieved from http://www.ocmboces.org/tfiles/folder889/Cell%20Phones%20in%20Schools%20Binder1.pdf

Fahnoe, C., & Mishra, P. (2013). Do 21st century learning environments support self-directed learning? Middle school students' response to an intentionally designed learning environment. In R. McBride & M. Searson (Eds.), *Proceedings of society for information technology & teacher education international conference 2013* (pp. 3131–3139). Chesapeake: AACE.

Johnson, L., Adams, S., & Haywood, K. (2011). *The NMC horizon report: 2011 K-12 edition*. Austin: The New Media Consortium Retrieved from http://www.nmc.org/pdf/2011-Horizon-Report-K12.pdf.

Knowles, M. (1975). *Self-directed learning: A guide for learners and teachers*. Englewood Cliffs: Prentice Hall/Cambridge.

Liu, M., Scordino, R., Geurtz, R., Navarrete, C., Ko, Y. J., & Lim, M. H. (2014). A look at research on mobile learning in K-12 education from 2007 to present. *Journal of Research on Technology in Education, 46*(4), 325–372.

Lloyd, J. (2010). *Cell phones head to class*. Retrieved from http://www.mysanantonio.com/news/education/article/Cell-phones-head-to-class-780555.php

Mitchell, L. (2014). Rise of the YouTube makeup artist: Nearly half of women use online tutorials. *Express*. Retrieved from http://www.express.co.uk/life-style/style/464633/Pixiwoo-Lauren-Curtis-Frozen-Nearly-half-of-women-copy-popular-online-makeup-tutorials

O'Bannon, B., & Thomas, K. (2015). Mobile phones in the classroom: Preservice teachers answer the call. *Computers & Education, 85*, 110–122.

Pew Research Center. (2015). *Teens, social media & technology overview 2015*. Retrieved from http://www.pewInternet.org/2015/04/09/teens-social-media-technology-2015/

Prensky, M. (2007, July–August). Changing paradigms. *Educational Technology*. Retrieved from http://www.marcprensky.com/writing/Prensky-ChangingParadigms-01-EdTech.pdf

Quillen, I. (2010). Schools open doors to students' mobile devices. *Digital directions*. Retrieved from http://www.edweek.org/dd/articles/2010/10/20/01mobile.h04.html

Saldaña, J. (2013). *The coding manual of qualitative researchers* (2nd ed.). Los Angeles: Sage.

Schenker, L. (2009). Schools embrace txt msg. *The Salt Lake Tribune*. Retrieved on September 24, 2014 from: http://www.sltrib.com/education/ci_12179870

Shuler, C. (2009). *Pockets of potential*. Retrieved from http://www.joanganz cooneycenter.org/wp-content/uploads/2010/03/pockets_of_potential_1_pdf

Teo, T. (2013). An initial development and validation of a digital natives assessment scale (DNAS). *Computers & Education, 67*, 51–57.

Teo, T., Tan, S. C., Lee, C. B., Chai, C. S., Koh, J. H. L., Chen, W. L., & Cheah, H. M. (2010). The self-directed learning with technology scale (SDLTS) for young students: An initial development and validation. *Computers & Education, 55*(4), 1764–1771. https://doi.org/10.1016/j.compedu.2010.08.001.

Part VII
Synopsis

Chapter 18
A Synoptic View

David Barlex and P John Williams

Abstract So, what are we to make of the findings embedded in each of the theses considered in this book? We organised the chapters into six groupings that we thought were of particular relevance to those who teach design and technology in schools:

- Curriculum content
- Stories of technology
- Planning and pedagogy
- Cognition
- Girls and technology
- Information technology

First this chapter will consider the findings of each grouping with a focus on their implications for classroom practice. Then the chapter will discuss the further work that might stem from the initial research and how teachers might contribute to this.

Findings Concerning Curriculum Content

Abbad Almutari's findings concerning primary school technology teachers in New Zealand and Saudi Arabia are interesting because they highlight a significant professional difference. In New Zealand where technology is a school subject in its own right, the teachers had a positive attitude to the subject and were able to use a variety of teaching methods which moved outside traditional instruction. In Saudi Arabia technology is integrated with science subjects and not treated as an

D. Barlex (✉)
University of Exeter Associate, Brighton, UK
e-mail: david.barlex@btinternet.com

P J. Williams
Curtin University, Perth, Australia
e-mail: pjohn.williams@curtin.edu.au

P J. Williams, D. Barlex (eds.), *Explorations in Technology Education Research*, Contemporary Issues in Technology Education,
https://doi.org/10.1007/978-981-13-3010-0_18

independent entity. The teachers did not consider that schools were ready to teach technology as a separate subject in its own right and were much more conservative in their approach to teaching. The integration of technology with science is fraught with difficulties for technology. The nature of scientific activity is significantly different from that of technological activity. Scientists observe and explore natural phenomenon with a view to developing explanations, whereas technologists intervene in both the natural and made worlds with a view to meeting peoples' needs and wants. Integrating science and technology in the school curriculum can lead to these differences becoming obscured with the result that teachers have a poor understanding of the nature of technology which, in turn, leads to a lack of professional confidence.

Victor Ruele explored how to enable teachers to grapple with change, such as the introduction of a new subject within the curriculum, in the context of secondary school technology education in Botswana. His findings are relevant to the situations in New Zealand and Saudi Arabia. His model for managing change involved six features, which were a shared vision amongst stakeholders, a strong and credible coalition between stakeholders, the creation of an enabling environment, provision of resources, continuous monitoring and professional development of those involved in the change process. If any were missing, the chance of the change being successful was reduced; and the more that were missing, the greater the chance of failure. In the case of New Zealand, most if not all of the conditions for change were met, whereas in Saudi Arabia, they were not, and in particular the establishment of a shared and unique vision for technology education unentangled with science education.

Eva Bjorkholm's research in Sweden focused on young children's learning in relation to understanding technical solutions to simple mechanical problems. The teachers of these children saw technology as an opportunity for practical work without much in the way of a specific focus on what was to be learned. Eva was able to identify four different 'ways of knowing' what was to be learned and from these was able to identify what needed to be taught for the children to evaluate the fitness of purpose of technical solutions. This detailed knowledge of learning intentions is essential if teachers are to have the professional confidence to teach technology effectively as a separate subject. Developing such knowledge is difficult if the nature of the subject is unclear.

Kieran McGeown explored secondary schools pupils' perceptions of practical work in the technology and design curriculum of Northern Ireland. He found that students saw some practical work as of limited value and uninterested in theoretical learning but that this changed when the practical and the theory were taught within an industrial context to which the students could relate. This points to an interesting difference from science. The phenomenon that is observed and explored in science is to a large extent independent of particular contexts as science tries to develop general explanations across the varying contexts in which the phenomenon is operating. Hence to be true to the nature of technology, it is important that the teaching and learning activities are placed in relevant contexts.

Findings Concerning Stories of Technology

Cecelia Axel analysed six Swedish children's storybooks, from the early 1900s to the early 2000s, that considered the nature of technology. She identified six themes and noted the presence of technology in nature in the past, present and future. Cecilia noted that the presence of the technology in the books often conveys an ambiguous message: on the one hand, the technology is fascinating, essential and a result of creativity. On the other hand, the technology is something that adversely affects both human relationships and nature. Hence teachers can use stories to explore what Arthur (2009) describes as our bipolar relationship with technology in terms of our trust for the natural compared to our suspicion of the artificial.

> These two views, that technology is a thing directing our lives, and simultaneously a thing blessedly serving our lives are simultaneously valid. But together they cause unease, an ongoing tension, that plays out in our attitudes to technology and in the politics that surround it. (page 214)

> …we trust nature, not technology. And yet we look to technology to take care of our future – we hope in technology. So we hope in something we do not quite trust. (page 215)

Sachin Datt considers the characteristic features of narratives that carry the cultural values of technology. He exemplifies this through the story of the Wright brothers and Otto and Gustav Lilienthal the innovators largely responsible for flying and controlling the first heavier than air machines. He identified 22 common elements of the stories but warned against using stories in which the inventors apparently derive their ideas 'in a flash of intuition' without reference to previous work by others. It is also worth noting that many innovations are the result of the collaboration between members of large teams although the outcomes of the teamwork are often seen as the 'brainchild' of a particular individual. Examples would include Steve Jobs of Apple; Craig Venter, responsible in large part for sequencing the human genome and developing synthetic organisms; and Elon Musk, developer of Paypal, the Tesla electric car and SpaceX to enable human space flight. Whilst the stories of these dynamic and charismatic individuals are appropriate for technology lessons revealing to a considerable extent the nature of technology and its underlying values, the role of others in the success of these endeavours should not be underestimated.

Findings Concerning Planning and Pedagogy

Eva Hartell explored the wide range of assessment issues that confront primary school teachers in the technology classroom in Sweden. She concluded that it was important to embed five key strategies into the teaching process for formative assessment to be successful:

- Clarifying and sharing learning intentions and criteria for success
- Engineering effective classroom discussions, questions and learning tasks

- Providing feedback that progresses learners
- Activating students as owners of their own learning
- Activating students as instructional resources for one another

Each of these is a demanding requirement in its own right; taken together they can become overwhelming for the teacher. Hence it is important that teachers are supported in this endeavour such that they can develop the self-efficacy necessary to build these strategies into their lessons in ways that are fluent and effective.

Mary Southall explored the way teachers in England, teaching pupils aged 11–14 years, planned teaching, learning and assessment for design and technology lessons. For the seven schools Mary investigated, she found that all teachers were required to use a whole school standardised planning pro forma whatever subject was being taught but that the design and technology teachers overlaid on this 'one-size-fits-all' approach a range of informal inputs that were often the result of serendipity. From a consideration of the planning scrutinised and lessons observed, Mary found that the approaches were generally limited with teacher talk and instruction dominating the lessons. Mary concluded that planning needed to be more focused on developing a learning environment in which the students might adopt a 'just-in-time' approach to the acquisition of knowledge and skills in response to learning activities. It should be noted that since Mary's research the design and technology curriculum in England has changed considerably with a greater emphasis on the acquisition of specified knowledge and the removal of attainment targets, procedural in nature, for monitoring progress (DfE 2013).

Tom Delahunty explored the way 3rd year undergraduates training to be teachers used problem-solving skills in tackling convergent tasks in which there was a single correct solution. He found that once a participant had conceptualised a problem, the approach to solving the problem was dependent on that conceptualisation. This is in contrast to expert problem-solvers who are able to be flexible in their approach and use different and multiple methods. Tom's concern was that many of the problems that young people tackle at school are of the 'single, correct solution' type and that being successful in such tasks is unlikely to prepare students for tackling open tasks with many possible valid solutions. For him the findings were particularly worrying as they came from students training to be teachers. Devising tasks for school technology lessons that are sufficiently open as to require flexible problem-solving but not so open that they are impractical for the classroom and overwhelm the learners is difficult, but technology teachers should not shy away from this challenge. Whilst convergent tasks have a place in helping to learn particular specifics, they need to be supplemented with divergent tasks.

Jason Power explored the effect of task difficulty on self-efficacy using the playing of Pacman configured with different levels of difficulty and reward. He found that self-efficacy could be manipulated by difficulty. If the task, in this case the game, was too difficult, then the player performed poorly and his/her self-efficacy diminished. Teachers are aware of the importance of task difficulty and how this should be matched to students' ability to tackle the task. If the challenge is too high, anxiety can result; if too low boredom sets in. Jason's research is relevant here, and he makes the important points that self-efficacy is enhanced through scaffolding tasks such that students experience mastery and feedback describing how individuals have improved.

Findings Concerning Cognition

Greg Strimel explored the way high school engineering students tackled an engineering design task and was able to identify the cognitive strategies they used and the extent to which they used them. The task was to design an easy to assemble filtration unit for reducing the turbidity of cloudy water. Although the students spent most time on model/prototype construction, analysing and managing, he found that those students who produced the best performing solutions spent more time on testing, experimenting, observing and interpreting data than those who produced poor solutions. As a result Greg recommends that the teaching of engineering design should involve specific instruction in the 'little used' activities and that students should be encouraged to adopt them even when, as is often the case, they simply want to construct the solution making ad hoc adjustments en route.

Michael Grubbs also looked at the cognitive strategies used by high school students. In this case the students were tasked with designing but not making a *Reach-n-Grab* device for a wheelchair-bound person to be able to reach and grab the glass jar when it is about 2 feet beyond their normal reach. Of the 40 students, 20 had taken a pre-engineering programme and 20 had not. Michael found that there was no difference in the cognitive strategies employed by the two groups of students. Taking the pre-engineering programme did not prepare the students for engaging with the design problem. This is perhaps not entirely surprising as technology and engineering programmes in the USA are electives with students choosing such programmes on a semester or annual basis with the possibilities of repeating or not repeating the programme and joining the programme in different school years. This makes the development of any long-term progression in design ability difficult. As a result of his research, Michael recommends that in developing learning activities for technology or engineering programmes one should consider how such activities will develop the cognitive strategies required for designing. His findings are very much in line with those of Greg Strimel in that explicit instruction with regard to strategies for designing is necessary if students are to become adept at making design decisions.

Findings Concerning Girls and Technology

Sonja Niiranen's research focused on how to increase girls' interest and motivation in technology education. Her findings from a questionnaire answered by over 600 fifth and sixth grade students in Sweden indicated that the social interaction between the teacher and the class was of particular importance to girls. From her interviews with women who were studying at university or working in technology-related fields, she found that significant others had played a part in their childhood with regard to providing significant intellectual capital that influenced their occupational choices. This does point to the difficulty of raising girls' interest and motivation as far as schools are concerned. It is difficult for teachers to become 'significant others' although some do manage this. And as indicated by the ASPIRES Project (ASPIRES 2014), for many girls the habitus of their family situation is such that they see occupations in technology-related field as 'not for people like us'.

Tackling the same issue in Australia, Vicki Knopke tried to identify the factors that encouraged senior female students at secondary school to participate and engage in technology education. Echoing the findings of Sonja Niiranen, Vicki found that the nature and range of learning styles in the classroom needed to reflect at least in part females ways of learning. Such learning styles whilst requiring structure also needed provide individuals with independence to solve problems and discuss issues surrounding the activities. One could add that such learning styles might well be relevant to males as it has been reported that male students often have tunnel vision with regard to tackling tasks, focus on the technical and avoid exploring wider issues (Murrphy D&T 4NG). In any technology classroom, there will be multiple stakeholders with varying aspirations and different needs and wants. The research by Sonja and Vicke indicate that it is the responsibility of the teacher to create a classroom ecology where such individual differences can be accommodated such that all young people, boys and girls, those likely to enter a technical profession and those not, find technology an interesting and worthwhile subject.

Findings Concerning Information Technology

Adrian O'Connor explored the use of a virtual learning environment (VLE) to support the teaching of 8 pre-service teachers and the learning of 104 pupils aged 15–18 across 7 classrooms in Ireland. The task undertaken by the pupils was to develop an improved version of an existing assistive technology. Adrian did much more than simply introduce the participants to the VLE. He helped the teachers organise lessons in accord with a model of the learning environment that considered educational transactions in terms of three factors – the experiential domain, the procedural domain and the individual domain – hence termed the EPI model. The result was that the teachers and pupils used VLE to support discourse, between the teacher and pupils and between pupils such that the learning process became much more explicit and the design ideas developed were much more robust than in previous situations in which the VLE was not available. This indicates strongly that the successful use of IT in teaching and learning requires that the pedagogy being used is sound in the first place and the IT supports and enhances this pedagogy. In this case the IT was able to facilitate enhanced discourse between teachers and pupils.

Scott Bartholomew explored the impact of the use of mobile devices on the way middle school students in the USA tackled the engineering design challenge of devising a pill dispenser. Did the use of such devices enhance the students' abilities to be self-directed? Scott found that access to mobile devices did not have a significant impact of student self-directed learning and that access to computers was negatively correlated with student self-directed learning. This has resonance with the findings of Adrian O'Connor in that access to IT alone does not necessarily enhance learning, in this case the ability to be self-directed. The message from both studies is that IT will only be successful in improving teaching and learning if it is utilised within effective pedagogy.

Suggestions Concerning Curriculum Content

The combination of science with technology in the school curriculum, a situation that pertains in Saudi Arabia, has the inherent difficulty that it might not be possible for both subjects to retain their identity. Abbad Almutari's finding that teachers in Saudi Arabia did not consider that schools were ready to teach technology as a separate subject in its own right indicates that further research into teachers' understandings of technology and how this might be enhanced to the point where they knew enough to be confident about teaching technology as a subject in its own right would pay dividends. Clearly teachers would have a key role to play in such research, not simply as 'subjects for investigation' but as co-collaborators with HE-based researchers in exploring the nature of technology and how this might inform a technology curriculum that was significantly and legitimately different from a science curriculum. This sort of research is strongly related to Victor Ruele's finding with regard to the conditions necessary to make curriculum change. In particular the shared vision and coalition between stakeholders are crucial, and establishing this would be an important first step for any initiative developing a new technology curriculum whether it be part of an integrated science and technology programme of study or as an entity in its own right. Enabling teachers to be a key stakeholder with a significant voice would be essential, and their role in each of Victor's identified conditions for change would enable them to make key contributions. It is important that these collaborative activities between HE and teachers in enabling curriculum change are disseminated widely amongst the participating stakeholder groups. Hence whilst members of the HE community will publish their findings in academic journals, the members of the teaching community should be supported and encouraged to write about their contributions in professional association journals. Joint presentations about this work from teachers and academics at education conferences would be useful in enabling those involved in technology education to meet with others in the wider education community.

Eva Bjorholm's findings focused on a particular aspect of young children's understanding of mechanisms in the wider context of technology education being required to teach specific technical content as well as engaging the learner in practical activity. These findings relate strongly to those of both Abbad and Victor in that any curriculum change must take into account the detail of the learning that is to take place in the classroom. This is where, so to speak, 'the rubber hits the road', and the only people in a strong position to know about and reflect on this are teachers. Hence teachers' reflections on the minutiae of their practice in teaching particular aspects of a technology curriculum are 'gold dust' in terms of educational research. Hence the collaboration between teachers and those working in HE can provide insight into practice and be used to enable those involved in ITE and professional development to introduce a wide range of teachers to these insights.

Kieran McGeown's exploration of pupils' perceptions of practical work highlighted the importance of listening to the views of young people about their learning experiences. The student is probably the most neglected of stakeholders in any attempt to introduce a new curriculum or change a curriculum, yet their voice can provide significant insights. A key question that might be asked of young people after they have finished a significant piece of work is: 'What would be your advice to the pupils who will be learning this topic next year?' One of the editors, David, asked pupils this question when he was piloting the Nuffield Design and Technology Project. The answers were always highly illuminating, sometimes beginning with the phrase, 'Well if I were the teacher I would…'. So an important role for teachers is to provide pupils with the opportunity to comment on the way they have been taught. It is important that this is not tokenistic and that the suggestions made are discussed and in some cases lead to changes in practice. Tackling this listening to pupils as a research activity in collaboration with those working in HE would provide fascinating insights into the world of the student as learner and could lead to both academic and professional association journal publications. Such studies could certainly form the basis for an MA assignment in technology education.

Suggestions Concerning Stories of Technology

Cecelia Axel's study has indicated that children's fiction can carry both positive and negative images of technology. Hence teachers might consider using fiction to help their students develop technological perspective (Barlex et al. 2015). The fiction concerned should be age appropriate and might even embrace media other than books. The way students respond to such fiction would make an interesting research study. It would be necessary for teachers to prepare the ground for the reading and help students find out, through discussion, what the fictions were saying about technology. In some cases it will be obvious that they will be cautionary tales, and the students' views might well relate to the concerns identified by Macnaghten et al. (2010) when they investigated people's attitude to nanotechnology:

- Be careful what you wish for – the narrative of desire
- Being kept in the dark – the narrative of alienation
- Messing with nature – the narrative of the sacred

- Pandora's box – the narrative of evil and hope
- The rich get richer – the narrative of exploitation

Barlex (2017) used these when considering a critique of robotics through science fiction films. They could provide the basis for an analysis of the students' considerations. Tackling this as a research activity in collaboration with those working in HE would provide fascinating insights into the way students might develop technological perspective and could lead to both academic and professional association journal publications. Such studies could certainly form the basis for an MA assignment in technology education.

Sachin Datt's idea of using a narrative to tell the story of individuals who are associated with a particular technological endeavour could be extended by teachers requiring their students to imagine what a technologically famous individual might do next. This would require helping students find out about the work of particular individuals and write under the title 'If I was XXXXX, what I'd do that I haven't done yet or didn't have time to do while I was alive would be…'. Such individuals might include figures from the past, Leonardo da Vinci and Alice Lovelace, or from the present, Elon Musk or Barbara Liskov. Tackling this as a research activity in collaboration with those working in HE would provide insight into students' views on the impact of individuals in the way technology works. Such studies could certainly form the basis for an MA assignment in technology education.

Suggestions Concerning Planning and Pedagogy

Planning learning so that appropriate teaching methods are used and the learning achieved can be identified by reliable assessment techniques is one of the most demanding aspects of teaching. Once a curriculum is established and appears to be successful, there is an understandable reluctance to change for the sake of change. But it is important that teachers reflect on the decisions that have led to the curriculum they are implementing, consider what might have influenced these decisions and decide whether change is required. Both the justification of existing practice and the identification for the need for change are important. Each of the researchers provided insight into different important aspects of this work and as such can be used as the basis for teacher reflection. Such reflection is best done in collaboration with a colleague as this enables a conversation about 'what we do?', 'why we do it the way we do?', 'in what ways is this successful and unsuccessful?' and 'what might we change for the better?' These are tough questions, and thinking about them on one's own is unlikely to be as successful as thinking about them with a colleague. Some of Eva Hartnell's research points to the need to consider student autonomy. This is echoed by Mary Southall's work which identified the need for students to be more proactive learners. Tom Delahunty's research pointed to the need for more open tasks, whilst Jason Power's research indicated the importance of scaffolding to enable students to achieve self-efficacy. Each of these features can be seen as lenses to use when engaging in collaborative reflective practice. It will be important to keep track of the reflections and make a record of the thoughts and decisions made in ways that are not onerous or over time consuming. The contents of such reflective diaries can inform the planning and pedagogy taking place in a department. They could certainly form the basis for an MA assignment in technology education.

Suggestions Concerning Cognition

Allowing students to use items such as craft knives, pillar drills, sewing machines or soldering irons without direct instruction as to their safe and effective use would be both dangerous and foolish. However, direct instruction alone is insufficient; students have to experience for themselves how to use such items, as the learning is, necessarily, to a large extent tacit. It relies on practice and the development of muscle memory. Both Gregg Strimel and Michael Grubb argue that there is a clear place for direct instruction with regard to the teaching of designing. It is similarly tacit in that it 'has to be done' to be learned although there are a range of strategies that can be taught that the student can use for designing. And it is here that teacher can explore which strategies to teach and the impact of these on students' designing. This is an interesting interplay between conceptual knowledge, knowing and understanding particular strategies, and procedural knowledge, knowing how to use particular strategies appropriately and effectively. Such an exploration would be interesting to other teachers and would easily find publication in the journals of technology teacher professional associations. Working with an HEI, the explorations of teachers from several schools could be combined to produce interesting academic papers that will be useful for technology teachers.

Suggestions Concerning Girls and Technology

The finding of both Sonia Niiranjen and Vicki Knopke indicates the importance of technology teachers developing and maintaining the classroom ambience that enables all students both boys and girls to engage fully and positively with technology education. This is particularly important with regard to girls as research such as that carried out by Sonja and Vicki indicate that in some classroom girls are marginalised. This is often unintentional, but nonetheless it has a deleterious impact on the way girls view themselves as able to develop either technological capability or technological perspective. So there is an important opportunity for teachers to explore their classroom environment with a view to understanding more about how this affects the way girls relate to the subject. This exploration is probably best done through paired observation in which teachers collaborate in observing each other's practice. It will be important that this is not carried out in an overly critical manner but in a way that simply observes and records what is happening leading to a joint conversation that questions this in a way that focuses on the way different students respond to particular features in a lesson and how the lesson might be adapted to enable a more positive response. This is no easy task and requires those involved to be both honest and sensitive in their search for a more inclusive curriculum. The process itself and the actions taken as a result could certainly form the basis for an MA assignment in technology education.

Suggestions Concerning Information Technology

Technology education finds itself in a situation where it can use information technology for two distinct purposes. Firstly information technology can provide software which students can use for both designing and making (CADCAM), and secondly it can provide software that teachers can use with students to enhance their learning such as making available a VLE as described by Adrian O'Connor or a mobile device as described by Scott Bartholomew. It is important not to confuse these uses of information technology. The findings of both researchers indicate that the use of information technology to enhance learning will only be successful if the underlying pedagogy is effective. There is an interesting opportunity here for teachers in developing curricula in which any information technology introduced to enhance learning is considered in the light of both what is being taught and how it is being taught. Will the information technology improve access to what is being taught? Is the information technology likely to be supportive of the pedagogy being used? The answers to these questions should be 'yes' if the information technology is going to enhance learning. The proof will be in the pudding so it will be important to scrutinise the resulting curriculum in comparison with the curricula that did not use any information technology enhancement. The findings will provide a guide as to whether the use of the information technology does make the desired improvement to student learning. This will be useful and interesting to other teachers and would easily find publication in the journals of technology teacher professional associations. Such an investigation could certainly form the basis for an MA assignment in technology education.

End Note

Whilst the doctoral-related research reflected in this book is just a snapshot of research taking place in technology education throughout the world, it does give an indication of the strength and diversity of the developing profession. Broadcasting this research, and making it useful for technology education professionals, is the goal of this book. To summarise, the research carried out by the contributors to this book is clearly of value, and teachers can certainly contribute to taking the research further. The teacher professional associations are in a position to support teachers in this endeavour through their own publications and conferences. University departments of education can support teachers on their MA programmes to carry out and report research activities that make further contributions to the fields explored by the PhD studies.

References

Arthur, W. B. (2009). *The nature of technology*. London: Allen Lane.

ASPIRES. (2014). http://www.kcl.ac.uk/sspp/departments/education/research/aspires/ASPIRES publications.aspx

Barlex, D. (2017). Disruptive technologies. In P. John Williams & K. Stables (Eds.), *Critique in design and technology education*. Singapore: Springer.

Barlex, D., Givens, N., & Steeg, T. (2015). Thinking about disruptive technologies. In G. Owen-Jackson (Ed.), *Learning to teach design and technology in the secondary school* (3rd ed.). Oxford: Routledge.

Department for Education. (2013). *The national curriculum in England framework document*. London: DfE.

Macnaghten, P., Davies, S., & Kearnes, M. (2010). Narrative and public engagement: Some findings from the DEEPEN project. In R. von Schomberg & S. Davies (Eds.), *Understanding public debates on nanotechnologies options for framing public policy*. Luxemberg: European Union.

MIX
Papier aus verantwortungsvollen Quellen
Paper from responsible sources
FSC® C105338

If you have any concerns about our products,
you can contact us on
ProductSafety@springernature.com

In case Publisher is established outside the EU,
the EU authorized representative is:
Springer Nature Customer Service Center GmbH
Europaplatz 3, 69115 Heidelberg, Germany

Printed by Libri Plureos GmbH
in Hamburg, Germany